T0360564

Cambridge Elements ≡

Elements in the Philosophy of Biology
edited by
Grant Ramsey
KU Leuven
Michael Ruse
Florida State University

CULTURAL SELECTION

Tim Lewens
University of Cambridge

Shaftesbury Road, Cambridge CB2 8EA, United Kingdom

One Liberty Plaza, 20th Floor, New York, NY 10006, USA

477 Williamstown Road, Port Melbourne, VIC 3207, Australia

314–321, 3rd Floor, Plot 3, Splendor Forum, Jasola District Centre,
New Delhi – 110025, India

103 Penang Road, #05–06/07, Visioncrest Commercial, Singapore 238467

Cambridge University Press is part of Cambridge University Press & Assessment,
a department of the University of Cambridge.

We share the University's mission to contribute to society through the pursuit of
education, learning and research at the highest international levels of excellence.

www.cambridge.org
Information on this title: www.cambridge.org/9781009539067

DOI: 10.1017/9781009539043

When citing this work, please include a reference to the DOI 10.1017/9781009539043

First published 2024

A catalogue record for this publication is available from the British Library.

ISBN 978-1-009-53906-7 Hardback
ISBN 978-1-009-53909-8 Paperback
ISSN 2515-1126 (online)
ISSN 2515-1118 (print)

Cultural Selection

Elements in the Philosophy of Biology

DOI: 10.1017/9781009539043
First published online: June 2024

Tim Lewens
University of Cambridge

Author for correspondence: Tim Lewens, tml1000@cam.ac.uk

Abstract: Humans learn in ways that are influenced by others. As a result, cultural items of many types are elaborated over time in ways that build on the achievements of previous generations. Culture therefore shows a pattern of descent with modification reminiscent of Darwinian evolution. This raises the question of whether cultural selection – a mechanism akin to natural selection, albeit working when learned items are passed from demonstrators to observers – can explain how various practices are refined over time. This Element argues that cultural selection is not necessary for the explanation of cultural adaptation; it shows how to build hybrid explanations that draw on aspects of cultural selection and cultural attraction theory; it shows how cultural reproduction makes problems for highly formalised approaches to cultural selection; and it uses a case study to demonstrate the importance of human agency for cumulative cultural adaptation.

Keywords: cultural selection, cultural attraction, Price Equation, cultural evolution, cultural adaptation

ISBNs: 9781009539067 (HB), 9781009539098 (PB), 9781009539043 (OC)
ISSNs: 2515-1126 (online), 2515-1118 (print)

Contents

Introduction: Culture from the Side of Natural History   1

1 The Arguments for Cultural Selection   5

2 The Attractions of Cultural Selection   19

3 The Cultural Price Equation   32

4 Waiting for Casabe   48

References   65

Introduction: Culture from the Side of Natural History

Charles Darwin's (1871) *The Descent of Man* set two quite different precedents for what it might mean to offer a natural historical perspective on change and stasis in human cultures. First, Darwin defended a conception of natural selection as *substrate-neutral* (Dennett 1995; Mesoudi 2011: viii). That is, he took the view that selection could be characterised in an abstract way such that it could range over entities of very different types: plants and animals of course, but also linguistic items like words. He also argued that selection could range over entities at different levels of organisation: individual organisms, but also communities of organisms.

When individual organisms differ in their abilities to confront the struggle for existence, the long-run effect can be the preservation and adding up of chance favourable variations which, given enough time, can give rise to the most exquisite adaptations. But individual organisms, says Darwin, are not the only entities that differ in their abilities to succeed under conditions of struggle: 'The survival or preservation of certain favoured words in the struggle for existence is natural selection' (1871: 60–61). Here, Darwin was endorsing the view of German linguist Max Müller, whom he quoted with approval in *Descent*:

> A struggle for life is constantly going on amongst the words and grammatical forms in each language. The better, the shorter, the easier forms are constantly gaining the upper hand, and they owe their success to their own inherent virtue. (1871: 60)

Darwin does not stop with words. If individual organisms can undergo selection, so can groups of organisms, and more specifically so can competing human communities: 'There can be no doubt that a tribe including many members who, from possessing in a high degree the spirit of patriotism, fidelity, obedience, courage, and sympathy, were always ready to give aid to each other and to sacrifice themselves for the common good, would be victorious over most other tribes; and this would be natural selection' (1871: 166).

Parental entities – whether words, animals or tribes – need to vary in their abilities to survive and reproduce, and their offspring must resemble them. But it does not matter *how* resemblance between parent and offspring is secured; forms of natural selection can occur just so long as resemblance *is* secured. This means that while natural selection at the level of human 'tribes' might be mediated partly by the inheritance of 'instincts', Darwin thought it could also be mediated by the ways those who are inspired by feelings of glory, 'excite the same wish for glory in other men, and would strengthen by exercise the noble feeling of admiration' (165).

Modern-day advocates of the notion of cultural selection – the notion that is the focus of this Element – have applied the substrate-neutral conception of natural selection with enthusiasm (e.g. Mesoudi et al. 2004, 2006). They have observed that whether one is dealing with technologies, scientific explanations, cooking skills or more or less anything else in the cultural domain, there is abundant variation. Tools are tinkered with and modified, scientists conjecture alternative hypotheses, foods are prepared in different ways. There is also resemblance across cultural generations as individuals learn from each other about the proper way to make a weapon, about received scientific wisdom, or about how to bake. Together these can result in the modification and accumulation of favourable cultural traits as alternative tools, explanations and techniques compete for the attention of users. Ultimately, the advocates of cultural selection conjecture, this can underpin a cumulative process capable of giving rise to increasingly effective approaches in domains that range from the processing of otherwise toxic foods to the design of space-flight technologies.

What commentators note far less often is that although *Descent* certainly endorses a substrate-neutral conception of selection, the explanatory use Darwin makes of this notion is sparing even when he is accounting for cultural phenomena that have the 'cumulative' character that is of interest to many modern researchers. *Descent* thereby sets a second more subtle precedent, acting as an exemplar for how to approach cultural evolution in a way that makes selective use of cultural selection itself.

A significant proportion of *Descent* is devoted to one aspect of cultural change; namely an account of how and why 'the moral sense' emerged in our species (Lewens 2007). The 'moral sense' is simply Darwin's name for the conviction humans have that some actions are worthy of praise, others worthy of blame. He offers his account, as he puts it, 'from the side of natural history', in a way that he hopes will supplement the more abstract pronouncements of philosophers (1871: 71).

His account begins with a canonical form of natural selection – that is the form of selection that arises through individual organisms having different levels of success in terms of survival and reproduction – acting to shape the instincts of human progenitors. In this early phase, these instincts direct beneficent behaviour towards self and offspring. Darwin then invokes natural selection at the level of 'tribes' to explain how beneficence expands beyond the immediate family. But selection disappears when he moves on to discuss how accepted rules for conduct have become further improved over time. In this context 'improvement', for Darwin, simply means that these rules approximate more closely to the Christian principle 'As ye would that men should do to you, do ye to them likewise' (106). In his telling, language, acting in tandem with intelligence, make it possible for humans to formulate and broadcast explicit rules for good conduct, which in turn are based on collective experience of what works to augment others' well-being. Reasoning also prompts humans to extend the domain of sympathy beyond the boundaries put in place by natural selection acting on individuals and groups: 'As man advances in civilisation, the simplest reason would tell each individual that he ought to extend his social instincts and sympathies to all the members of the same nation, though personally unknown to him. The point being once reached, there is only an artificial barrier to prevent his sympathies extending to the men of all nations and races' (1871: 100–1).

This story is gradualist and (for the most part) cumulative. In a series of steps that build on those that have gone before, our ancestors move from motivation that is directed only at the well-being of their 'tribe' to that of all sentient beings, and they move from having only a shaky grip on what augments welfare to holding a better-informed view. Still, Darwin makes no effort to shoehorn this story into a selectionist idiom.

A form of cultural selection is present in Darwin's natural history of culture, but it is just one element of an approach that draws in opportunistic fashion on a whole variety of forms of learning, reasoning and communication. This Element does not make a case for any of the specifics of Darwin's approach, which is sometimes overtly racist, sexist and eugenic in character. But it does follow Darwin in supporting an eclectic account of cultural evolution, in which cultural selection is flexibly interpreted to suit explanatory needs, and in which selectionist approaches couple with other resources in the explanation of cumulative cultural change.

Here is how this Element proceeds. Section 1 notes the diversity of different background motivations that make the concept of cultural selection important for the theorists who use it. This explains why cultural selection is defined in different ways – some of which may seem peculiar on superficial inspection – by different theorists. There is no need to choose just one definition of cultural selection as the 'correct' one. But this does not mean that these definitional issues are of no importance: in particular, it is essential to distinguish between a loose view of the *pattern* of cultural change as one characterised by gradual modification of previous efficacious techniques; and various far more disciplined accounts of 'cultural selection' itself as a *process* by which this cultural accumulation takes place. On close examination, it turns out that many of the most prominent researchers in cultural evolution already embrace a view that sees cultural selection strictly defined as just one element – and not even a necessary one – of the more loosely defined set of quasi-selectionist processes by which cultural adaptation occurs.

Section 2 furthers this flexible and pluralist approach to cultural adaptation. It has become common in recent years to distinguish two dominant 'schools' of cultural evolutionary thinking – the 'Paris' and 'California' Schools – and to suggest that while the California School has a focus on both cultural selection and cultural adaptation, the Paris School instead minimises the importance of both in favour of their key resource of cultural 'attraction' (Sterelny 2017). I argue for a rapprochement that I believe to be implicit already in much of the work of both schools. Evolutionary developmental biology ('evo-devo') shows how concerns with biological adaptation are informed both by natural selection and by the ways in which the organisation of organisms facilitates some forms of variation (Gerhart and Kirschner 2007). Drawing on an analogy with evo-devo, I make a case for the complementarity of selectionist approaches often linked to the California School, and the Parisian notion of cultural 'attraction', in the explanation of cultural accumulation.

Section 3 steps back to consider a particularly rigorous approach to cultural selection that draws on the work of George Price (1970, 1972, 1995). I look at Price's own early discussion of the features a general account of selection needs to have, as well as more recent efforts to put forward specific formulations of the Price Equation that are suitable for application in the context of cultural evolution (El Mouden et al. 2013). This section shows difficulties the cultural domain raises for drawing the crucial distinction that the generalised Price approach requires between facts about an entity's productivity with respect to the subsequent generation, and facts about resemblance between parents and offspring. These difficulties are especially easy to see in the light of the approach to cultural attraction previously defended in Section 2.

The content of Sections 1–3 is highly abstract. This is hardly surprising, given that the very idea of cultural selection relies on an abstracted account of what selection is, as well as of the many different types of entities selection can range over. Section 4 closes the Element in more concrete form with a detailed case study of techniques used by Tukanoan people of Amazonia, which make the poisonous manioc tubers they cultivate safe to eat. I focus on manioc processing because it has become emblematic of the importance of a 'hidden-hand' approach to cultural adaptation, which parallels the hidden hand of natural selection in Darwin's work (Henrich 2016). In particular, I look for a rationale that explains the two-day waiting process before Tukanoan people deem the bread that they make from manioc – called 'casabe' – ready to cook. I argue that closer attention to the manioc case supports the eclectic approach this Element advocates: too much of a focus on 'blind' processes of cultural imitation renders the origination of effective processing techniques inexplicable. The advent of manioc processing can only be understood when due stress is also placed on human agency, on judicious learning, and on forms of cultural 'attraction' that draw on the general taste for fermented foods.

There are many issues of central relevance to cultural selection that are not addressed in this Element. There is next to no discussion of cultural group selection, a key explanatory concept for many leading theorists of cultural evolution (e.g. Henrich 2004; Richerson et al. 2016; Heyes 2018). That is not because I think cultural group selection is unimportant: on the contrary, it is because a proper discussion of it brings so many complexities that it could fill an Element in its own right. The same goes for the topic of 'Universal Darwinism', the wholly general view that all instances of good 'fit' between an entity and its environment must necessarily be explained by some sort of selection process, working at some level of organisation or another (Dawkins 1983). I briefly discuss Universal Darwinism in Section 1.3, simply to note the background motivating role it plays for some (but not all) advocates of cultural selection. Universal Darwinism raises issues about processes of adaptation within minds and brains, within the developing bodies of plants and animals, and even within artificial neural networks, that go well beyond the scope of this Element. Finally, space does not allow for a full evaluation of the voluminous psychological evidence relevant to cultural adaptation, especially detailed evidence characterising the ways in which humans and animals learn from each other, and from their environments. Elements are required to be short, and my hope is that this one does enough in defending an eclectic approach to cultural adaptation.

1 The Arguments for Cultural Selection

1.1 Pattern and Process in Cultural Evolution

Many secondary school children in the UK know things about the Universe that neither Charles Darwin, nor even Isaac Newton, were able to figure out. They know, for example, that the cells in human bodies contain chromosomes, that chromosomes are composed of DNA molecules, and that DNA molecules have a double-helical structure.

Although they know these things, these children did not discover them. They were taught by others, who were taught the very same things in turn when they were at school. Sometimes people do discover new pieces of knowledge; but when they do this, they are always relying on more learning from others, because they are always putting even earlier discoveries to work. James Watson and Francis Crick were the first to propose a double-helical structure for DNA, but their suggestion relied on what they had learned from other scientists about crucial properties of the molecule. For example, work using

X-ray diffraction techniques – undertaken by Rosalind Franklin, Maurice Wilkins and others at King's College London – played an essential role in Crick and Watson's thinking (Cobb and Comfort 2023). And Franklin and Wilkins did not invent X-ray diffraction: it had been honed by previous generations of scientists such as Dorothy Hodgkin (who used the technique to uncover the structure of molecules involved in organic processes), and before her Lawrence and William Henry Bragg (who developed ways of discovering the structures of simple crystals from the patterns cast by X-rays that passed through them), and before them Max von Laue (who showed that a crystal would scatter X-rays by diffraction).

Scientific research offers just one illustration of the ways in which human achievements in one generation rest on work done by those who came before. Peter Richerson and Robert Boyd instead encourage their readers to think about 'being plunked down on an Arctic beach with a pile of driftwood and seal skins and trying to make a kayak' (2005: 130). Someone like me would probably fail (I grew up on a farm in the South-West of England), but Inuit people would be more likely to succeed. The difference does not lie in their possession of innate knowledge of kayak manufacture, and nor does it lie in each individual Inuit being creative and resourceful enough to discover from scratch how the kayak construction process needs to go. Unlike the DNA example, this case concerns knowledge of how to do something, instead of knowledge that something is the case. But like the DNA case, what the Inuit know is learned from others, and it has been consolidated and augmented over time.

These two untheorised examples help to illustrate the appeal of the notion of cultural evolution. In both cases there is an impressive end-product, built by gradual accumulation: knowledge of the basic structure of a significant biomolecule, or knowledge of how to build a kayak. Just as Darwin (1859) suggested that the many functional adaptations found in the natural world are modified versions of earlier structures inherited from ancestors, so the techniques, technologies and bodies of knowledge that enable us to understand, manipulate and thrive in the world are also modified versions of techniques, technologies and bodies of knowledge that have been passed to us from earlier generations (Basalla 1988).

There are important questions about how fine-grained the resemblances are between the patterns of change in the cultural and biological domains. Although both areas allow lines of descent to be traced over time, and although change in both areas is often cumulative and gradual, there are many issues of detail about the pace of change, and about the extent to which change in either domain can be represented as a branching tree. It is obvious that in the cultural context a given artefact can incorporate elements drawn from many different technological traditions; and it is less obvious to what extent these cross-lineage borrowings also characterise organic evolution, and how cultural evolutionary researchers should deal with the problems they pose (Gray et al. 2007).

Setting these fine-grained questions of *pattern* to one side, the important question for this Element is whether it is possible to tell a selectionist story about *processes* of adaptation in both spheres. Darwin did not merely argue that the adaptations observed in the natural world are modified structures inherited from ancestors. He also gave an account of how exactly this gradual adaptation occurs. His answer appeals in part (but only in part) to natural selection. This occurs, in his view, (i) whenever parents differ in their abilities to confront the struggle for existence; (ii) whenever offspring resemble parents with respect to the traits that confer success in this struggle; and (iii) whenever new variation can be introduced to these changing lineages. Over time, says Darwin, occasional favourable variations will be introduced by chance, they will spread because of their positive effects on the reproductive output of parents, and they will then serve as the bases for further beneficial variations as and when they arise. It will become clear in Sections 2.4 and 2.5 that the conditions for natural selection to act can be in place without any guarantee that complex adaptations will appear. But even if natural selection does not constitute a complete explanation, Darwin showed how it could be an important part of the explanation for cumulative adaptive change.

In Sections 1.2–1.4, I give a brief overview of three different starting points that have led theorists to argue that the process of adaptation is similar within the cultural and organic domains, and to build their cases for the importance of a form of cultural selection accordingly. Section 1.5 pauses the investigation of cultural selection itself, to note the role of that concept within the much broader context of cultural evolutionary theory. Section 1.6 then surveys some of the many ways in which cultural selection has been spelled out in detail. The point of this section is not to argue that just one of these definitions is the right one. It is to explain why cultural selection is defined in different ways, by appealing to differing motivations for putting the notion to work. In Section 1.7, I show that some of the most prominent theorists of cultural evolution are best understood as advocating quasi-selectionist approaches. These approaches borrow aspects of natural selection thinking in the explanation of cumulative adaptation, but they do not amount to cultural selection in a strict sense. But before any of that, I outline three different pathways that have all led theorists to endorse a notion of cultural selection.

1.2 The Recipe-First Approach

First, and perhaps most obvious, what I call (borrowing from Godfrey-Smith 2009) the *recipe-first approach* argues that the general recipe required for natural selection is instantiated in the cultural domain. Natural selection requires that parents differ with respect to reproductive output, and that off-spring resemble parents with respect to traits that confer reproductive success. Likewise, there are lots of different ways of making a kayak, with some working better than others. Observers can see which construction techniques give rise to better boats, and adopt the techniques that they prefer. Techniques can thus be said to reproduce, in the sense that they reappear in the hands of learners in ways that resemble their enactment by demonstrators. These propensities to repro-duce vary within populations, giving rise to a form of cultural selection. This approach is exemplified by several prominent theorists of cultural evolution. Mesoudi, Whiten and Laland (2004: 1), for example, have argued that 'cultural evolution has key Darwinian properties', and they have in mind the conditions required for selection, which they understand to be 'variation, competition, inheritance, and the accumulation of successive cultural modifications over time'.

1.3 Universal Darwinism

A second perspective, which I call the *Universal Darwinism approach*, differs in the logical strength of the appeal it makes to selection. The recipe-first view says that as a matter of fact the conditions required for selection are instantiated in the cultural realm. The Universal Darwinism approach says that selection is the only process that can *possibly* give rise to instances of good fit between some entity and the demands placed upon it (Dawkins 1983; Cziko 1997). 'Good fit' is used in an expansive manner here, to capture the full range of circumstances whereby entities are tuned to match their goals. It is intended to cover the ways in which organisms are adapted to their environments, the ways techniques are adapted to their ends, the ways tools are well-suited to their purposes, and even the ways theories capture the domains they are meant to represent. It follows from Universal Darwinism that if adaptation is observed in the domain of culture, some form of selection must be responsible for it as a matter of necessity.

The psychologist Donald Campbell is perhaps the most significant defender of this view. He argued that: 'A blind-variation-and-selective-retention process is fundamental to all inductive achievements, to all genuine increases in know-ledge, to all increases in fit of systems to their environment .... In such processes there are three essentials: (a) Mechanisms for introducing variation;

(b) consistent selection processes; and (c) mechanisms for preserving and/or propagating the selected variations' (Campbell 1974: 421).

This view is compatible with the recipe-first approach, and indeed the two positions are often held at once. Even so, they are not identical. If Darwinian selection is the only process capable of giving rise to adaptation, and if significant adaptations can be found in the cultural domain, then it follows that these adaptations must be explained by some selection process or another. That process might turn out to be selection acting at the level of cultural items such as techniques, in a manner that complements the sketch I gave in Section 1.2 relating to changes in kayak design. But Universal Darwinism itself leaves open the question of the level at which selection processes occur. Instead of going on among techniques held within cultural groups, they might go on in the minds of clever innovators, or perhaps they go on across biological generations in ways that give rise to innate knowledge. So it is possible to sign up to Universal Darwinism, while dismissing cultural selection as a process of little or no importance. Steven Pinker (1997) seems to have a view of this sort. Conversely, the recipe-first approach is compatible with a denial of Universal Darwinism, for one might think that while culture is a domain where important selection processes occur, there are also non-selectionist processes that can account for the emergence of 'good fit'.

I cautioned in the introduction that an evaluation of Universal Darwinism is well outside the scope of this Element. That said, it is worth drawing attention to one apparent disanalogy between Darwin's conception of natural selection, and the forms of change that go on in the cultural realm, that might seem fatal both to Universal Darwinism and to the recipe-first approach. Alternative kayak designs are not produced entirely randomly: innovators do not throw together any old combination of skin, bone, wood, leaves and so forth before waiting to see what works best. Instead, their novel designs can be guided by significant prior knowledge of what is, and what is not, likely to work well on the water. In brief, cultural adaptation draws on guided variation, natural selection does not, or so some critics of cultural selection have claimed (Pinker 1997: 49).

A full evaluation of this supposed disanalogy raises a whole series of difficult questions about exactly what is meant by saying that variation for Darwin is random, whether it is true even in the biological world that variation is always un-directed or un-guided, what the relationship is between the notions of 'random-ness' and 'guidedness', and so forth (see among many others Kronfeldner 2007 and Jablonka and Lamb 2014 for valuable discussions). In lieu of attempting to answer those questions, let me instead clarify that advocates of the recipe-first view fully admit that variation in the cultural domain is often guided in some sense: 'Selection occurs *anytime* there is heritable variation that affects survival or reproduction . . . . It does not matter whether the variation is random' (Henrich, Boyd and Richerson 2008: 129; see also Mesoudi 2008). Whether this is a threat to the explanatory force of cultural selection depends on how pervasive this guidedness is, and also on what questions the investigator is asking in the first place (Amundson 1989; Chellappoo 2022). If one is trying to understand the origination – rather than the diffusion – of innovation, then cultural selection will have no explanatory role to play if designers are insightful enough to single-handedly invent the steam-engine, or the personal computer, in a perfect form that requires no further modification. No inventor is so insightful, and (as I elaborate further in Section 2.5) the selection-driven proliferation of *moderately* well-designed gadgets throughout a population of innovators can be part of the explanation for how designs emerge that are better still, even if designers take those earlier partial successes and tinker with them in intelligent ways, using whatever fallible insights into engineering principles they can muster.

Universal Darwinists like Campbell (1960) also acknowledge that sometimes the generation of good fit draws on prior knowledge. Designers use rules of thumb and other heuristics to ensure that they do not even bother trialling designs that would likely be utterly hopeless. Campbell points to a need to explain where these rules of thumb and heuristics come from – how, that is, does one know that there is no point trying to use particular materials, techniques or configurations in putting together a new form for a kayak – and his answer is that at some point in history they must have been the result of a process that was truly 'blind': 'The many processes which shortcut a more full blind-variation-and-selective-retention process are in themselves inductive achievements, containing wisdom about the environment achieved originally by blind variation and selective retention' (1960: 380). Selection does not disappear here: it is simply moved into the explanatory background.

1.4 The Case-Based Approach

There is a third perspective, which I call the *case-based approach*, that also motivates appeals to cultural selection. This pragmatic perspective does not begin from any general commitment to the notion that selection is the only process that can explain adaptation; nor does it begin from the abstract observation that cultural processes satisfy the conditions required for natural selection. Instead, its proponents point to specific instances of highly effective behaviours, tools or even institutions used by humans to cope with their environments and with each other. For example, Joseph Henrich highlights how many communities have developed extremely effective ways of processing foods to remove toxins (2016). Drawing on details of those cases, he suggests that the individuals who use the techniques often have little or no idea of what makes them so efficacious, and yet they (rightly) persist with them all the same. He argues (in ways that I will discuss in much greater depth in Section 4) that the best-supported explanation for the origination and preservation of these techniques is that they are selectively copied from predecessors in ways that allow for gradual improvement over time.

The case-based approach is once again compatible with the other two, but it is not the same. Beginning with specific cases, and the explanatory puzzles they pose, is potentially a good way to highlight why it matters that the general recipe for selection is instantiated in the cultural domain. However, the case-based approach also allows for cultural selection as such to be accorded a restricted role, in favour of a more eclectic approach to cultural adaptation. I will argue in Section 1.7 that this is exactly the approach that Henrich's detailed work points to.

1.5 Cultural Selection and Cultural Evolution

So far, I have explained some of the different motivations that explain why various theorists have placed weight on cultural selection; but I have not said how cultural selection is understood in any detail. The three different background motivations help to explain why that notion is spelled out in such different ways in works on cultural evolution. In a moment I will give a sense of how diverse these understandings are. Before doing so, it is important to take a step back, and to situate the notion of cultural selection within the wider project of cultural evolution (Lewens 2015).

'Cultural Evolutionary Theory' names a programme of research that draws on disciplines that include archaeology, cognitive science, anthropology, economics and evolutionary biology.[1] Proponents of this approach describe it as 'cultural' because of its focus on the ways in which humans – and other organisms – learn from other members of their species. These abilities are often described as forms of 'social learning', and sometimes as constituting a channel of 'cultural inheritance'. Cultural evolutionary theorists tend to ask questions about the nature and origins of these capacities for social learning, and also about the effects of social learning on how populations change and adapt over time. The answers to these questions are mutually informing, because a focus on the effects of particular types of learning can feed into subsequent explanations for the benefits these dispositions bring, and the reasons why they can become stabilised in populations.

---

[1] The next two paragraphs are adapted from Lewens and Buskell (2023) and Lewens (2020).

I have explained what makes cultural evolutionary theory cultural. What makes it evolutionary is complex, but some relevant factors include the following:

i.   Researchers in this tradition often examine how learning interacts with the forms of inheritance (especially genetic inheritance) studied by mainstream evolutionary theorists.

ii. They often draw attention to learning as a means by which humans and other animals engage adaptively with their environments, often placing particular stress on the stepwise manner in which forms of learned adaptation can improve over generations.

iii. They seek to understand culture using explanatory models and investigative tools adapted from those used in evolutionary and ecological theory.

iv. They reach back into human pre-history when determining the origins of the capacity for culture.

v.   They ask comparative questions concerning differences between species in terms of their abilities to create and maintain storehouses of valuable socially transmitted information.

This broad approach to Cultural Evolutionary Theory is most thoroughly exemplified in the ongoing research tradition initiated in the 1970s and early 1980s by Cavalli-Sforza and Feldman (1981) on the one hand, and Boyd and Richerson (1985) on the other.

All the most prominent evolutionary theorists of culture have on occasions made use of notions of cultural selection. But not every question they raise requires an answer in terms of cultural selection. For example, some of Marcus Feldman's earliest work on cultural inheritance, undertaken in collaboration with Cavalli-Sforza, was dedicated to undermining overly simple (and highly misleading) inferences from claims about populational distributions of IQ scores to claims about genetic causation (Cavalli-Sforza and Feldman 1973). Their models pointed to ways in which learning from common familial environments could give rise to a spurious appearance of a strong role for genetic variation in accounting for differences between families: 'Given the existence of individual plasticity in response to the environment, correlations between biological relatives are expected *even if there is no genetic variation whatsoever*' (1973: 633, emphasis in original). This aspect of their work does not draw on cultural selection, and more generally it does not concern cultural adaptation.

Importantly, even when cultural evolutionary theorists are focused on the ways in which learning enables individuals to adapt to changing environments, it is still not the case that they always draw heavily on notions of cultural selection. For Henrich, one of the central goals of cultural evolutionary research is to understand how and why our species became able to construct increasingly elaborate and effective techniques and technologies:

> Probably over a million years ago, members of our evolutionary lineage began learning from each other in such a way that culture became cumulative. That is, hunting practices, tool-making skills, tracking knowledge and edible-plant knowledge began to improve and aggregate—by learning from others —so that one generation could build on and hone the skills and know-how gleaned from the previous generation. (2016: 3)

Henrich accounts for this capacity by leaning on specific claims about how humans learn. For example, he argues for the importance of copying the actions of others in detail, even when the purposes of many steps in a complex sequence are opaque to both learner and demonstrator. He also argues that humans have evolved strategies that lead them to copy particular types of people – the successful, the prestigious, the healthy – when it is too difficult to figure out more directly which techniques would yield the biggest payoffs if they were imitated (more on both of these themes in Sections 4.4–4.6). Henrich puts far more weight on specific ways in which (in his view) accumulation happens, than on making a general case for cultural selection as the engine of this stepwise adaptation.

1.6 The Varieties of Cultural Selection

There is more, then, to cultural evolution than cultural selection. But cultural selection is the notion that interests me in this Element, and it is time to briefly review some of the many ways in which the term is used.[2] Within mainstream evolutionary biology, natural selection is always understood to range over populations in which entities differ in fitness. Even here a plurality of definitions ensues, because there is a proliferation of answers given to the questions of (i) what entities constitute the relevant populations (e.g. individual organisms, genes, groups of organisms, even species), and (ii) what their 'fitness' consists in (e.g. their actual number of immediate offspring, their total causal contribution to the production of offspring within some group, their probabilistic disposition to produce offspring down many generations, their mere ability to persist across time regardless of offspring produced, etc. etc.). It is not surprising, then, that if one tries to give an account of cultural selection, one is faced with a similar range of problems. Are the bearers of cultural 'fitness' (i) publicly observable tools, techniques or behaviours; (ii) internal mental states of some kind, such as ideas, representations or beliefs; (iii) people who can act as

---

[2] This section is adapted from Lewens (2022).

demonstrators; or something else entirely? And does fitness consist in, to give just a few options, (i) the disposition of a technique to be copied, (ii) the rate at which some type of idea spreads through a population, or (iii) the ability of an individual person to attract learners?

This abundance of apparently reasonable ways of spelling out what cultural selection is does not constitute a problem, so long as theorists are explicit enough in their work to avoid confusions. For example, close to the birth of formalised cultural evolutionary theory, Cavalli-Sforza and Feldman noted that learned traits can straightforwardly affect what they call *Darwinian fitness*: an individual organism's chances of surviving and reproducing can be affected by what it learns, and in some cases the trait in question may learned by the organism's offspring, too. They distinguished this form of natural selection mediated by social learning from what they called *cultural selection*, which they defined 'on the basis of the rate or probability that a given innovation, skill, type, trait, or specific cultural activity or object—all of which we shall call, for brevity, *traits*—will be accepted in a given time unit by an individual representative of the population' (Cavalli-Sforza and Feldman 1981: 15). A culturally inherited trait might be highly detrimental to the Darwinian fitness of an organism, and yet spread very rapidly through cultural selection (as they understand that term). For example, one can imagine a seductive form of religious celibacy that completely effaces the organism's Darwinian reproductive fitness, while nonetheless being appealing enough to others that it is widely adopted. Conversely, it is possible to imagine skills in the preparation of medicines which are (i) valuable with respect to survival and (ii) so hard to learn that only biological offspring of experts have the time and opportunities to acquire them. Such traits might augment Darwinian fitness, while enjoying low levels of cultural selection.

Much more recently a similar distinction, albeit with different labels, has been used by Birch (2017) and Birch and Heyes (2021). They use 'Cultural Selection 1' (CS_1) to refer to a process whereby cultural variants proliferate because they cause their bearers to have more offspring (who in turn inherit those traits by learning them). This corresponds precisely to Cavalli-Sforza and Feldman's selection based on Darwinian fitness. However, Birch and Heyes's 'Cultural Selection 2' (CS_2) does not quite match what Cavalli-Sforza and Feldman call cultural selection. That is because cultural fitness in Birch and Heyes's CS_2 is measured by the number of learners a trait causes individuals to attract, rather than by the general rate of acceptance of the trait.

To see why these two ways of understanding cultural fitness do not match, imagine a technique of some kind that is blindingly obvious, but which only becomes useful if individuals need to use computers. An example might be some way of pressing multiple keys at once. As computers become widespread in a community, this technique might see a very swift increase in acceptance in a population – and be 'fit' in that sense – simply because everyone starts to figure it out for themselves. This swift rate of proliferation may have nothing to do with the individuals who use the technique attracting lots of learners. This example satisfies the definition for Cavalli-Sforza and Feldman's 'cultural selection', at least if one agrees that the rate of acceptance of the technique by representative individuals within the population increases per unit of time, but it does not satisfy the definition for Birch and Heyes's 'CS_2'.

Sometimes theorists have understood cultural selection in ways that are surprising, and may wrongfoot those who expect cultural fitness to simply reflect a trait's propensity to spread in a population. Consider Richerson and Boyd's (2005) mode of distinguishing between what they call cultural selection (which corresponds to Birch and Heyes's CS_2), and what they instead refer to as biased transmission (which has no obvious counterpart in Birch and Heyes's taxonomy). They know very well that their definitional practice is likely to appear peculiar, but they have good grounds for their choice. They justify it by reflecting on how the notions of selection and biased transmission are distinguished in organic evolution (see also Henrich et al. 2007).

An example will illustrate their reasoning. Suppose that because of differences in their abilities to outrun predators, slow-running deer have fewer offspring than fast-running deer. Suppose, also, that slow-running deer have slow-running offspring, and fast-running deer have fast-running offspring. If a population begins with a 50/50 split between slow and fast runners, then in the next generation the proportion of fast runners will increase. This is a change brought about by natural selection. Now suppose, instead, that predators are absent, with the result that slow runners and fast runners have just the same numbers of offspring. Suppose, also, that because of a quirk of genetic inheritance (this could be due to a phenomenon such as meiotic drive, but details do not matter here) both fast runners and slow runners tend to have fast-running offspring. This time, when we look to the next generation, the population will again change so that the number of fast runners increases. But this is not usually understood in evolutionary theory as a result of selection: even though the fast-running trait increases its frequency, that is not because fast-running individuals are fitter on average than slow-running individuals. This change occurs instead because of a bias in the process of inheritance that affects how the traits in question are transmitted to the next generation.

Boyd and Richerson transpose this distinction into the domain of culture in the following way: when an individual attracts more learners than others because they possess a given cultural trait, this is cultural selection. But when different cultural variants have different chances of being transmitted to a learner, we are once again dealing with transmission bias rather than selection.

This proliferation of definitions is to be expected: investigators have choices over exactly how they define cultural fitness, and those choices necessitate corresponding adjustments in how cultural selection is understood. Nature itself is complex enough that it does not recommend a single definitional option as obviously the best. Nonetheless, Richerson and Boyd's account of cultural selection leads to results that will seem surprising to those who think cultural selection happens whenever there are repeated instances of improvements to techniques or technologies as they spread through populations. Suppose there are two different methods of making a kayak, A and B. Suppose, also, that individuals using A and B are just as likely to attract learners: assume, for the sake of argument, that the people who use A and B do not differ in their social standing, credibility and such like, so that learners are just as likely to pay attention to a model who uses A as they are to pay attention to a model who uses B. Even so, imagine the learners in question find B much easier to understand and remember than A. The result is that B spreads much more effectively through the population. It might be that B is easier to acquire because it does the job in a much more sensible way than A, and so the steps needed to execute it are more intuitive and therefore more memorable. In other words, B is simply a better way of making a kayak. This means that B could end up spreading because of transmission bias – and more specifically because of what Richerson and Boyd call 'content bias' – rather than because of what Richerson and Boyd think of as selection. Moreover, transmission bias could be iterated in such a way that it leads to B's replacement by an even better method C, C's replacement by D, and so forth. When understood in Boyd and Richerson's very specific way, cultural selection is not necessary for cumulative cultural adaptation (Lewens 2015).

1.7 Cultural Adaptation without Cultural Selection

I suggested in Section 1.1 that differences in how cultural selection is spelled out can derive from the different background motivations that prompt theories of cultural selection in the first place. For the Universal Darwinist, for example, selection processes are necessary for adaptation. Processes of cultural change are processes whereby adaptations – understood as traits showing good 'fit', such as effective kayak designs – arise. It follows that the Universal Darwinist must opt for a definition of selection capacious enough to encompass these cumulative processes of cultural change. This may be why Alex Mesoudi tends to think of phenomena of content-bias (whereby cultural variants are chosen for copying based on utility, attractiveness and so forth) as forms of cultural selection (2011: 65). The very fact that iterated choices based on increasing utility can give rise to the accumulation of beneficial variations means this has to be understood as a form of selection for the Universal Darwinist, and it drives the Universal Darwinist to understand selection in terms of the differential spread of cultural variants in a population. Moreover, there is a rationalisation of this choice within easy reach: cultural selection, like natural selection, can be understood as a process whereby variants that have a better match with functional demands laid down by their environments (which, in this case, means the environments constituted by the demands of users) are more likely to proliferate. Meanwhile, for those who are unwedded to the notion that only selection can explain adaptation, nothing is lost by insisting on a strict distinction between cultural selection and biased transmission, and assimilating cases of content-bias to the latter.

Henrich's approach to cultural selection differs once more. He notes, 'Like natural selection, our cultural learning abilities give rise to "dumb" processes that can, operating over generations, produce practices that are smarter than any individual or even group' (2016: 12). But although these processing may be 'like' natural selection because they are 'dumb', Henrich does not refer to them as processes of cultural selection. In fact, in his book-length overview of the cultural evolutionary project, Henrich does not use the term 'cultural selection' at any point. I do not mean to over-stress the significance of this fact: *The Secret of Our Success* is written for a broad market, and Henrich does address cultural selection (especially, but not only, at the level of groups) in his many more technical works (e.g. Henrich and Boyd 2002; Henrich 2004). Even so, he is motivated to understand the human capacity for cumulative culture, and to do so using whatever explanatory resources he can lay his hands on. Thus, he tends to write of the ways in which 'cultural evolution—through . . . selective attention and learning processes . . . is fully capable of generating . . . complex adaptive processes which no one designed or had a causal model of' (2016: 114).

Henrich's approach can be described as *quasi-selectionist*, because he often highlights relevant similarities between natural selection and these 'selective attention and learning processes'. This does not, however, amount to a defence of a general notion of cultural selection in a strict sense. Instead, he is alluding to a series of more specific, testable claims about how (in his view) individual learners are disposed to focus their efforts on particular kinds of people, and on particular kinds of behaviours. These discriminating learning tendencies arise (according to Henrich) as old-fashioned natural selection favours individuals whose dispositions to learn are the more effective in harnessing information from others, and who have more babies because of that: 'Individuals whose genes have endowed them with the brains and developmental processes that permit them to most effectively acquire, store, and organize cultural information will be the most likely to survive, find mates, and leave progeny' (2016: 64). To give just one example of what Henrich has in mind, he argues that individuals are likely to be better off if they show an exaggerated tendency to copy what the majority around them do, and that this explains the evolution by natural selection of so-called 'conformist bias'.

Henrich's rough reasoning can be illustrated using a simple example. Imagine you are a novice gardener wondering what the most effective way to grow tomatoes is. And suppose it is the beginning of the season, so you cannot look directly at the success of others. To make things easy, suppose you have a choice between growing your tomatoes: (a) in a greenhouse all the time; (b) in a greenhouse to start, and then outdoors; or (c) outdoors all the time. Finally, suppose that as a matter of fact the best approach is to use a greenhouse all the time. If others in your community have learned how to grow tomatoes by trial and error, then it is unlikely that they will all have arrived at this best approach. But even if others in the population are highly fallible, the majority approach can still offer a good signal of the best technique. Hence, there may be good information to be had by paying attention to what everyone else is doing and following the most popular choice.

I will assume that Henrich is right about all this, and that people do have an exaggerated tendency to imitate the majority (see Lewens 2015 and Morin 2016 for caveats). His fundamental interest is to show how conformists have greater biological fitness than nonconformists, because they are better able to exploit information in their social environments. It is perfectly reasonable to describe conformist bias as a form of 'selective learning': individuals show a discriminating disposition to copy what majorities, rather than minorities, are up to. In these cases of selective learning, individual learners are literally 'selecting' whom to learn from. This offers a reversed perspective on cultural change compared with notions of cultural fitness, which instead make cultural selection a consequence of the differing dispositions of demonstrators (or perhaps the traits they hold) to be copied. In the first case, 'selection' is a matter of which demonstrator a specific individual chooses; in the second case, 'selection' is a matter of which type of demonstrator, or perhaps which type of trait, has the greater success in terms of attracting learners, or spreading through a population.

There are ways of shifting from the perspective of individual selections by learners to the perspective of cultural fitness and cultural selection. One could say that individuals who happen to hold the majority view have a higher cultural fitness, simply because others are more likely to attend to them; alternatively, one could say that traits that are widely held are ipso-facto fitter, regardless of any further details of what those traits are. There would be some justification for these choices, because evolutionary biologists also sometimes link a trait's fitness to its frequency in the population, as when a disposition to fight for resources pays off when most others are timid, but it becomes a liability when most others are willing to fight. The key point here is not that Henrich's approach to conformist bias makes it impossible to talk about cultural fitness (and see Ramsey and De Block 2017 for a defence of this notion). The key point, instead, is that there is no urgency within this type of project – which primarily asks about the biological fitness consequences of learning dispositions – to fashion and deploy a notion of cultural fitness.

Henrich and colleagues summarise their stance in an early work noting that, 'We believe that constructing a full-fledged theory of cultural evolution requires considering a longish list of psychological, social, and ecological processes that interact to generate the differential "fitness" of cultural variants' (Henrich et al. 2007: 129). Their decision to hold 'fitness' within quotation marks here suggests that their research programme places far less weight on securing the importance of a form of cultural selection than it does on marshalling an eclectic range of explanatory resources that show how learning dispositions originate, how they need to be characterised, and how their interactions give rise to cultural accumulation. In Section 2, I will show how such an eclectic theory can draw on the resources of 'cultural attraction'.

2 The Attractions of Cultural Selection

2.1 Cultural Evolution in California and Paris

It has become common in recent years to distinguish two broad 'Schools' of cultural evolutionary thinking: the California School and the Paris School (Sterelny 2017). In very broad terms, what distinguishes the Californians from the Parisians is that the former place far more stress than the latter on explaining cultural adaptation, and also on forms of cultural selection as resources for understanding this phenomenon. For the Parisians the explanatory focus is instead on the existence of stable traditions (whether adaptive or not), and the key explanatory resource they put to work (which I introduce in Section 2.3) is the notion of cultural attraction (Acerbi and Mesoudi 2015; Sterelny 2017).

Up until now, this Element has dealt almost exclusively with the California School, so called because of the key roles played by collaborators Peter

Richerson (of the University of California at Davis) and Robert Boyd (whose PhD was also undertaken at UC Davis, and who worked for many years at the University of California Los Angeles). They have trained many other prominent cultural evolutionists such as Richard McElreath and Joseph Henrich, and their (1985) *Culture and the Evolutionary Process* has become an exemplar for the field in terms of questions to be asked and methods to use in addressing them. The 'California School' label also encompasses an equally important pair of founders of the modern discipline of cultural evolutionary thinking. They are Marcus Feldman and Luigi Luca Cavalli-Sforza, both of California's Stanford University, and authors of the (1981) *Cultural Transmission and Evolution*. Feldman is an exceptionally prolific scientist, and he has also had a hand in training many leading researchers of a new generation within cultural evolution, such as Laurel Fogarty and Nicole Creanza. Meanwhile, the 'Paris School' names the group of researchers in cognitive science and cultural anthropology who have grown up around Dan Sperber, of the Institut Jean Nicod in Paris. They include Olivier Morin, Christophe Heintz, Hugo Mercier and Nicolas Claidière.

The Paris/California binary should not be taken too seriously. There are many extremely prominent researchers who do not fit neatly into either 'School': Cecilia Heyes, Kevin Lala, Alex Mesoudi and John Odling-Smee, for example, were all trained in one way or another by Henry Plotkin of University College London, and so one could also attempt to baptise a third 'London School'. Some researchers have a claim to an affiliation with both Paris and California: Olivier Morin, for example, belongs to the Paris School if anyone does. Yet Morin is also co-author of a paper on cultural selection (El Mouden et al. 2013) with distinctly Californian themes. These labels are caricatures, but they are useful all the same for sketching some differences in approaches to cultural evolution.

2.2 California Revisited

I have already given a characterisation of the California School (albeit not under that label) in Section 1, but a recap is in order. Recall that Richerson and Boyd, and also Henrich, often like to make their case for an evolutionary approach to culture by spelling out the ways in which learning from others gives rise to impressive suites of techniques that help people to survive in harsh environments. It is learning from others, for example, that explains how techniques evolved that have enabled Inuit people to build dwellings, to find food, and to clothe themselves in the polar winter. Boyd and Richerson have put the point more generally: 'The single most important adaptive feature of culture is that it allows the gradual, cumulative assembly of adaptations over many generations—adaptations that no single individual could invent on their own' (2000: 148). The same goes for Henrich, whose central concern is with 'The striking technologies that characterize our species, from the kayaks and compound bows used by hunter-gatherers to the antibiotics and airplanes of the modern world', and who makes a case that they 'emerge not from singular geniuses but from the flow and recombination of ideas, practices, lucky errors, and chance insights among interconnected minds and across generations' (2016: 5–6).

The production of cultural adaptation is by no means the only focus of these thinkers, but it is a prominent one. They do not invariably turn to cultural selection strictly defined when they explain cumulative adaptation, for reasons I gave in Sections 1.6 and 1.7. But they are *quasi-selectionist* in their approach, in the sense that they draw attention to similarities between the processes of cultural adaptation and natural selection. For example, Boyd and Richerson argue that what is required for 'cumulative adaptive evolution' in the cultural domain is that, 'culture constitutes a system maintaining heritable variation' (2000: 158). This focus on 'heritable variation' – which requires that advantageous techniques, or behaviours that appear in one generation can be preserved in the next – does not necessarily require accurate copying between individual people, but it does require some mechanism or another that achieves preservation at the level of the population. This is why, burrowing into the details, the Californians often point to learning dispositions such as 'prestige bias' (of which more in Section 4.6), or 'conformist bias' (Section 1.7), in an effort to explain how population-level preservation can be achieved in spite of error-prone learning at the individual level.

To give a flavour of this approach, consider a modern classic of the literature, namely Henrich and Boyd's (1998) formal model intended to illustrate how conformist bias – which I introduced in Section 1.7, defined as the exaggerated tendency of individuals to adopt the most common cultural trait in a population – can overcome the effects of error-prone learning to produce reliable inheritance at the population level. Their model shows how, in a population that already contains several different traits at significant frequencies, the effect of error on a population-wide distribution of traits is low, because different errors tend to balance each other out. In a population in which one trait is common, the effects of error are much more significant. But because conformist bias gives the majority trait a disproportionate influence over what individuals learn, this bias increases the chances of a commonly held trait remaining commonly held in future generations, even with error-prone imitation. Hence, (they argue) conformist bias means that the overall makeup of the later cultural generation is likely to resemble that of the earlier generation.

2.3 Cultural Attraction

The Paris School is best understood via their key concept of *cultural attraction*. Cultural attraction, in turn, is well illustrated by seeing how its originator – Dan Sperber – used it to criticise a very particular version of cultural evolutionary theory, namely memetics, or meme theory (Sperber 2000). In this Element, I dwell on memetics only for long enough to explain how Sperber has responded to it. That is because, with some notable exceptions such as significant works by Dennett (e.g. 1995, 2017), memetics has not had much influence among researchers on cultural evolution.

Meme theorists follow Richard Dawkins (1976) in claiming that evolution in general, whether cultural or biological, requires replicators. These are entities whose function is to make copies of themselves. This power of replication ensures that slight advantages can be preserved and further improved; moreover, some replicators are more successful than others in making copies of themselves, by virtue of the effects they exert upon their environments (including the internal environments of organisms, partly constituted by the effects of other replicators). Dawkins's *The Selfish Gene* explores, for the most part, the replicators that underpin (in his view) the old-fashioned natural selection that takes part in the organic world. These replicators are, of course, genes. But Dawkins also argues that there is a second type of replicator, and a potent one, to be found in culture. These are *memes*, and he gives a famous list of some exemplary ones: 'tunes, ideas, catch-phrases, clothes fashions, ways of making pots or of building arches' (1976: 192). Just as genes make copies of themselves with different levels of success, and come to dominate in or disappear from populations according to the downstream effects they exert on the organisms and environments in which they dwell, so ideas – perhaps competing ideas about religious observance, or left-wing politics, or the efficacy of vaccines – also make copies of themselves as they hop from mind to mind, and they too come to dominate in or disappear from the social groups in which they are found depending on their effects on the minds that house them. Or so the story goes.

What Sperber points out in his (2000) critique of the meme theory is that even when ideas and behaviours are broadly the same from one individual to the next, it need not be because they act as replicators. His claim relies on a disciplined understanding of what a replicator is. Take genes as the exemplary replicators: when they make copies of themselves, it is because of a precise matching process, whereby each base-pair is copied bit-by-bit to produce a resembling strand of DNA. Now, suppose that you have some notion of how to make a cake, and you act on it: I see your cake, and its manifest deliciousness prompts me to make a similar one. It may be that my cake resembles yours very closely indeed.

One explanation for this resemblance is that I attend very closely to your actions and copy them bit-by-bit. It might also be that I spend a lot of time with you, and you explain to me your conception of how the cake should be made: in this way, I might also succeed (albeit using the outwardly audible evidence of your words) to copy bit-by-bit your internal idea of how the cake should be made. But I may simply glance at your cake, and be prompted to reflect on how I might go about making an equally delicious one. I briefly think things through, put my own knowledge of cake-making to work, and infer how I might go about making a similar one. If my tastes and my training are similar to yours, then I might succeed in producing such a cake – I might even end up with a very similar internal idea to yours of how to make the cake – with no bit-by-bit copying. So a form of 'reproduction' – in the sense of a new production of something that is similar to what came before – occurs here, without 'replication' in any strict sense.

Humans are reflective, creative agents who attend to the information available to them, and process that evidence in ways that are guided by their background knowledge, emotional dispositions and so forth. What they do in response to the situation they are in may also be affected by such diverse factors as the raw materials accessible to them, the bodily movements that their anatomy and physiology make most easily available, and so forth. These remarks are all platitudes, but they serve as a reminder that behaviours and ideas can reliably reappear in many individuals across populations because of the biasing action of similar psychological and physiological dispositions being put to work in similar environments, rather than because of any effort at fine-grained copying. Sometimes these instances of reappearance rely on dispositions that are shared only at a local level: for example, it may be very easy to for employees to remember and repeat a particular in-joke, because it draws on shared perceptions of their boss, or of the company they work for. Sometimes reappearance may instead rely on dispositions that are shared far more widely: for example, grief may spread through a population at times of national mourning because of aspects of emotion that stretch across all humans. The overall position has been summarised by Scott-Philipps and collaborators thus: 'the cognitive mechanisms producing social transmission—most obviously those involved in communication, but others too—do not in general aim at high-fidelity copying as such . . . . Cultural stability emerges as the cumulative effect of many non-random (i.e. biased) transformations' (Scott-Phillips et al. 2018: 162).

When Sperber writes of cultural 'attractors', he is explicit that the term is not meant to point to any specific form of explanation. It is intended as an abstraction: an attractor is any type of cultural item – a form of behaviour, a technique an idea or belief – that reappears with moderate reliability not because of bit-by-bit copying, but because shared background resources give rise to similar effects across groups of people (Sperber 1996). The notion of attraction draws attention to a series of more concrete potential explanations. These explanations only become informative when specific 'factors of attraction' – namely the particular dispositions that underpin these resemblances in some context or another – are spelled out.

To see how this spelling-out works, consider a representative 'Parisian' paper by Olivier Morin (2013). My goal here is not to support or undermine Morin's analysis, but just to give a sense of the sort of explanatory pattern typical of the School. He has made a case that over the course of the Renaissance one sees an increase in the proportion of portraits in which the subject appears to stare directly at the viewer, as opposed to staring at an angle that does not meet the viewer's own gaze. Morin asks, 'Is the evolution of Renaissance portraits consistent with cognitive attraction?' (224); however, this misrepresents what he is aiming to do, because the framework of cognitive attraction is so capacious that it's hard to see how any specific episode could be inconsistent with it. Morin's paper is really dedicated to detailing some of the specific 'factors of attraction' that explain this transition to direct-gaze paintings.

At the very beginning of his writing on this topic he notes work in psychology indicating that research subjects rate direct-gaze images as more attractive; that these images grab the attention of subjects more effectively; and that even very young babies prefer to look at direct-gaze pictures. He then makes the case that many paintings have become famous because they are direct-gaze: the alternative (which he argues against) is that the fame of direct-gaze paintings is merely a side effect of the fame of the sitter or the painter. He also makes a case that a shift in the preferred style of painters underlies a shift to more direct-gaze portraits. Here, again, there is an alternative hypothesis: painters just paint whatever is in front of them, and over time the sitters prefer to look at the painter rather than look away. Putting all this together, Morin argues that a general psychological preference for direct-gaze paintings – one that living art critics have, as well as long-dead Italian painters, and very young children recruited to psychological experiments – is an explanatory factor (although certainly not the whole explanatory story) underlying the increased proportion of direct-gaze paintings during the Renaissance.

2.4 Cultural Attraction and Cultural 'Evo-Devo'

Cultural attraction theory looks to map the various ways – at different levels of scale from the whole human species to narrower communities – in which shared emotional responses, pieces of background knowledge, affordances of raw materials, even bodily dispositions, impart biases to the sort of cultural variation that is, and is not, likely to be generated in a population. Specific pieces of work on 'factors of attraction' undertaken by the Parisians have often pointed to psychological biases that they also believe to be both species-wide and innate. That said, none of these three features – that biasing factors reside in psychology, rather than affordances of the body or of materials used; that they are shared by all humans, rather than being specific to more restricted communities; or that they are innate, rather than the products of plastic developmental processes – is truly essential to the notion of cultural attraction. Whatever the scale, and whatever the developmental story, these biases on cultural variation can sometimes act in ways that help to generate stable traditions over time.

This means that so-called 'evolutionary-developmental biology' – 'evo-devo' for short – offers an obvious conceptual model, drawn from within mainstream evolutionary theory, for how cultural attraction theory might be combined fruitfully with cultural selection. The starting point for evo-devo is the observation that genetic mutation can only alter phenotypic traits by having some influence on the developmental systems of organisms. Hence the nature of those systems plays a significant role in accentuating some forms of change, and dampening others. In other words, biases on variation are stressed in the evo-devo programme, just as they are stressed by the Parisian school.

Some of the biases explored by evo-devo's proponents may be highly generic. The likes of Newman and Müller (2001) have regularly argued that attending to the underlying thermodynamic properties of the material components of cells helps to explain why some configurations of living matter are more readily attainable than others. But there can also be far more local biases specific to narrower taxonomic groups, and which are subject to change over time as developmental systems change. Once the basic form of explanation is grasped here – namely, that one discerns which forms of variation are encouraged and which are discouraged by developmental systems – then it becomes evident that the question of what developmental resources are available within taxonomic groups at various ranks helps to shed light on what range of variation is, and is not, facilitated (see Lewens 2009). A great many studies – whether they are in beetles, seals or butterflies – now highlight the fact that not all forms of variation are equally likely to arise (see Salazar-Ciudad and Jernvall 2010; Uller et al. 2018 among many others).

A role for biases in the forms of variation that can arise does not preclude forms of selection, whether natural or cultural. Hence, one finds within mainstream biology various efforts to integrate the study of biased variation with the study of selection and the generation of adaptation. If one understands the forms of variation that developmental systems make available to selection, then one can understand why adaptation proceeds in one direction rather than another. And, of course, these processes exert mutual effects on each other: as selection plays a role in shaping developmental systems, so it plays a further role in altering the forms of variation that are available for subsequent rounds of selection to promote or reduce.

Shifting back from mainstream biology to cultural evolution, the exemplar laid down by evo-devo research suggests there should be plenty of scope for the study of cultural attraction to contribute to the study of cultural adaptation even though, as a matter of fact, researchers on cultural attraction have tended to focus on explaining phenomena such as the stability of traditional folk-tales over time, thereby choosing examples that do not invite thoughts of cumulative improvement (e.g. Morin 2016). I here follow Mathieu Charbonneau (2016) in pointing to an evo-devo-inspired approach to culture. He has pointed out, for example, that one gains a better understanding of the cultural evolution of stone tools if one also understands how features of human musculature, properties of raw materials, and so forth mean that some pathways for change in tool configuration over time are readily accessible, and others are not.

My suggestion, then, is to see the theory of cultural attraction as a theory focusing on the diverse sets of facts that make a difference to the tendencies of communities at various levels of magnification to produce variation of one type, rather than another. This does justice to Sperber's initial (1996) intentions for cultural attraction. It also offers an account that is more restrictive than those offered by some other recent proponents of cultural attraction, including (at times) Sperber himself. For example, cultural attraction should not be used as a synonym for the much more general project of developing an 'epidemiology of representations', another term associated with the Paris School. As Sperber (2001) sees things, epidemics are population-level events that need to be explained by aggregating two kinds of event at the level of individuals; facts about individual pathology, and facts about social transmission. By analogy, an 'epidemiological' approach to culture is any approach that explains changing distributions of different types of cultural item in a population by reference to both individual psychology, and cultural transmission. Cultural attraction and cultural selection, as I am recommending the terms be used, are specific explanatory resources one can bring to these more general questions of cultural epidemiology. To repeat the point, the resources of cultural attraction help to explain which ideas, or techniques, are most readily made available as variants within a population. But this does not remove the potential relevance of a different set of notions – Richerson and Boyd's narrow notion of cultural selection, for example – that point to the ways in which some individuals may become far more prominent as demon-strators than others. This also means that I reject Scott-Philipps et al.'s suggestion that, 'cultural attraction is the probabilistic favoring of some types of items over others' (2018: 164), and Claidière, Scott-Philipps and Sperber's similar suggestion that an attractor is 'any type whose relative frequency tends to increase over time' (2014: 5). If cultural attraction occurs whenever some items are more likely to increase their frequency in a population than others, then cultural attraction becomes such a capacious term that it summarises all the underlying factors that might make a difference to a cultural trait's prospects, including factors relating to the prominence of models bearing the trait. It is more useful to reserve cultural attraction for a more limited set of explanatory resources (see Acerbi and Mesoudi 2015; Buskell 2017a, b).

2.5 Cultural Attraction and Cultural Adaptation

Joseph Henrich spells out a central contention of the California School: 'We can survive because, across generations, the selective processes of cultural evolution have assembled packages of cultural adaptations—including tools, practices, and techniques—that cannot be devised in a few years, even by a group of highly motivated and cooperative individuals' (2016: 27). As I have already indicated, it is important not to confuse a claim about pattern – namely, that many important cultural traits are the result of gradual modifi-cation and improvement over time – with a claim about the specific process that explains this.

I have also noted that the California School care far more about showing how groups can retain favourable variations, which can then be further built upon, than they do about showing that a form of cultural selection is responsible for cultural adaptation. Indeed, they explicitly show how these two processes can come apart: the lesson they draw from their modelling of conformist bias is that groups can have the ability to maintain distributions of variation even when the instances of one-to-one learning within those groups are highly fallible in terms of bringing about reproduction at that level. Under these circumstances, selec-tion at the level of groups is not what explains the emergence of adaptation, because there may only be one group in play, and selection at a given level requires a population of entities at that level. Selection at the level of individuals may not be what explains the emergence of adaptation either, because individ-uals may learn from so many others, and in such fallible ways, that it makes little sense to analyse these episodes using any notion of cultural fitness, or cultural selection, within the group.

In Section 1.7, I pointed out that the Californians often endorse a quasi-selectionist approach, which stresses the importance of the reliable reappearance of cultural patterns at the level of populations if cumulative adaptation is to occur. I also argued that quasi-selection should not be equated with a strict account of cultural selection. Even so, at one point Boyd and Richerson (2000) go beyond the mere idea that populations need to stably retain cultural patterns over time, suggesting that cultural 'traits' – remember this is a catch-all term for things like techniques, ideas, even behaviours – also need to demonstrate 'fecundity' if cultural evolution is to be cumulative. I think they are right to pick out a distinctive role played by fecundity, even in quasi-selectionist explanations of cultural adaptation.

The first step to seeing this is to review how selection explains adaptation in central biological cases.[3] The question is not merely how selection can move a fitness-enhancing trait already present at a low frequency in a population – a better functioning eye, say – to fixation. Instead, the point of interest is how well-functioning eyes come to exist at all. A view that has now become widely accepted states that selection explains the origination of adaptations by increasing the number of individuals with fitness-enhancing traits (e.g. Neander 1995; Lewens 2004). In other words, one should not merely credit selection with a 'sieving' role of weeding out less beneficial traits. Rather, by making some traits more prevalent within a population, the chances of new traits arising that are yet more beneficial are increased.

Selection can do this if structures ordered by fitness correspond to structures ordered by mutational likelihood. Suppose that a functional eye C is more easy to reach via mutation from a semi-functional eye B, than it is from no eye at all A. Then, in a population comprised primarily of A individuals with just a few Bs, the chances of a C mutation increase as selection increases the absolute number of the fitter B compared with A. There is no guarantee that things will work out like this: it might, after all, be easier to generate a functional eye C from no eye at all A, than from a semi-functional eye B. Under these circumstances, if selection makes B more prominent in a population because it is fitter than A, selection also reduces the chances of a further C mutation arising. In short, systems need to be appropriately organised if selection is to play the creative role accorded to it. Even so, selection can play this creative role, if those organisational requirements are met. This is another reason (once again amplifying messages stressed by Charbonneau) for attending to the processes affecting how variation is generated, and gives further incentive to integrate theories of attraction with selectionist theories of adaptation.

---

[3] The remainder of this section is adapted from Lewens (2022).

This role for selection in origin explanations – stressed by Godfrey-Smith (2012), Heyes (2018), myself (Lewens 2004), and others – focuses on the manner in which the chances of adaptation are increased by multiplying the (absolute) number of suitable 'platforms' for further improvement. It is possible for attraction to underpin the multiplication of platforms that characterises selection-based explanations for adaptation.

Suppose we focus on increasingly effective versions of a ship's compass (Boyd and Richerson 2000). Some individuals observe compass design A, others observe a superior design B. Suppose that those exposed to B are far more likely to adopt it than those exposed to A, because it is easier to learn. This assumes that B's superior functioning is also accompanied by an elegance or transparency in its design that make it more intuitive to reverse-engineer than A. This assumption certainly need not hold with generality – a design that works better may be more complex, harder to understand and harder to learn, than an inferior one – but it may be true sometimes. In other words, B may constitute an attractor. Learners do not need to copy it slavishly; instead, the fact that B is an intuitively better design means learners do not find it difficult to infer steps that suffice to produce it. Even so, if it is easier for someone to discover how to produce an even better compass C, if they already know the steps required to produce B, then the proliferation of B is part of the explanation for C's arrival in just the same way that selection explains adaptation via 'multiple platforms'.

In short, attraction is compatible with the basic selectionist schema for explaining adaptation (see Driscoll 2011 for much more detailed arguments that extend and complement this point). Note, also, that the case under consideration might be one where some individuals, in fact, observe very many loosely resembling B-type compasses, while others observe very many loosely resembling A-type compasses. The 'copies' produced might be broadly inspired by *all* the several tokens they engage with. Under these circumstances, one cannot strictly say that any one token compass (or token representation of a compass) is singly responsible for the production of a resembling token. Likewise, no single model is chosen as the basis for copying. So the notion of cultural 'fitness' again struggles to find a foothold, even when a quasi-selectionist form of explanation is in play.

Godfrey-Smith (2009: 154) has followed Sperber (1996, 2000) in suggesting that the 'machinery' required for there to be cultural 'lineages that can be described in terms of reproduction' only exists 'when the agents apply simple habits of imitation, picking behavioural models and copying them without transformation and customization of the behaviour acquired'. So in some cases there may be no 'reproduction' in this moderately demanding sense, plenty of 'attraction', and yet the explanatory aspect of the proliferation of multiple platforms, central to selectionist explanations of adaptation, is present all the same. This is a hybrid quasi-selectionist explanation of cultural adaptation that draws on ingredients taken from several different schools of thought.

Godfrey-Smith suggests that: 'In both biology and culture, successive rounds of undirected variation can yield significant design improvements, provided that the successful variants in one generation proliferate and provide many independent platforms at which further innovation can occur' (2012: 2166). The 'hybrid' case sketched earlier raised the possibility that some cases of cultural adaptation might be explained by a combination of cultural attraction stabilising the reappearance of functional tokens in a population, while proliferation of those tokens also helps to explain the increasing chances of even more functional tokens appearing in the future. Godfrey-Smith's contention is that only with such proliferation (whether underpinned by attraction or not) can one expect un-directed variation to produce adaptation.

To what extent is this constraint loosened if variation is partially directed? One can imagine cases where designs circulate among just a handful of talented and knowledgeable innovators, in ways that lead to gradual improvements over time. The innovators in question observe and tinker with the designs of others in this small group in creative and intelligent ways. (Again, both the stabilisation of designs within this community and the ability of individuals to ascertain further functional innovations might be underpinned by various factors of 'attraction'.) Such a situation might well be one where favourable design variations are selectively retained. In that very loose sense, it is a selection process, but it does not include the key feature of selection-based explanations of adaptation, namely the proliferation of many independent platforms for further adaptation. There may simply be insufficient absolute numbers to enable this statistical form of explanation to have much significance.

2.6 Cumulative Adaptation without Proliferation

This point about numbers is worth stressing. Bernard Carlson has shown how Thomas Edison used sketches to develop his telephone (Carlson 2000). In 1862, the German scientist and inventor Philipp Reis demonstrated an apparatus that he called the *telephon* to Wilhelm von Legat, Inspector of the Royal Prussian Telegraph Corps. The report Legat wrote on the Telephon was translated by Western Union (including the diagram reproduced in Figure 1), and given to Edison in 1875. Edison then quickly produced a series of 'notes', albeit in the form of another diagram (reproduced in Figure 2), in which he offered alternatives to Reis's design. The supplements included a floating rheostat, which was based on an earlier approach of his own, and the diagram also shows an

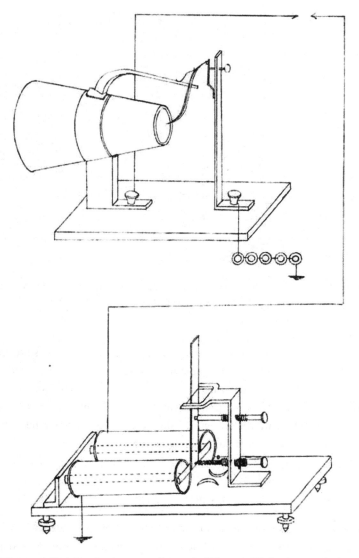

Figure 1 Reis's telephon, as shown in Western Union's translation of Legat's report.

alternative device for reproducing sound, which he labelled as using 'mercury like Helmholz'.

To be sure, this is a case of cultural descent with modification, but it is not a case that instantiates the general selectionist schema for cumulative improvement. It is also a case where one individual effects considerable steps forward by bringing together, in a re-interpretative way, a series of earlier innovations

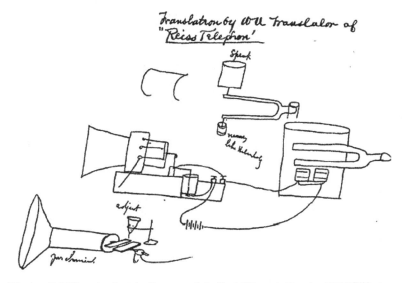

Figure 2 Edison's sketch diagram, labelled 'Translation by WU [Western Union] translator of "Reiss Telephon"'.

derived from his own earlier work, his contact with Helmholz, and so forth. The populational aspect – whereby multiple promising platforms are created across a wide group of individuals, all of which might serve as potential loci for further improvement – is entirely absent. In summary, while the selectionist schema can be integrated with cultural attraction, the selectionist schema need not be present in all explanations of cultural adaptation.

3 The Cultural Price Equation

3.1 What Price Culture?

Section 2 made a case for integrating selectionist forms of explanation with the notion of cultural attraction. It reinforced a more general message established in Section 1: cultural selection needs to be interpreted and deployed in a flexible way, with due attention to the explanatory goals it is being put to in a given context. There is a potential response to this proposal: perhaps a more disciplined, and less malleable, account of cultural selection could be fashioned by paying more attention to our best formal account of what selection, in general, is. In particular, one might suggest that the approach due to George Price offers by far the most compelling interpretation of cultural selection. In this section, my goal is to argue that a thoroughgoing effort to apply this formal approach to the cultural context shows how processes of central biological importance – selection in particular – do not always transfer in easy ways to the cultural realm. This is not because of well-documented worries (briefly mentioned in Section 1.3) about the supposed 'guidedness' of cultural variation. Instead, it is because forms of cultural reproduction pose more fundamental problems for drawing the distinction between selection and transmission that is central to the evolutionary understanding of selection itself.

I show this by reviewing how the 'Price Equation' can sometimes run into trouble when it is transferred from mainstream evolutionary theory to the cultural domain. These problems are most easily seen through the lens of cultural attraction itself. Here is how Section 3 proceeds. First, I introduce the Price Equation and explain its appeal. I then use Price's own reflections on generalised forms of selection to show how applications of the equation rely on a strict ability to distinguish (i) how productive an entity is with respect to offspring, from (ii) the degree to which an entity's offspring resemble it. The former is taken to measure the strength of selection, the latter instead corresponds to what is often called 'transmission bias'. In the context of the Price Equation, 'transmission bias' (introduced with the example of running speed in deer in Section 1.6) simply measures the extent to which offspring differ from their parents in some respect, and it takes a value of zero when they are identical. This measure is entirely neutral about whether the differences in question are explained by quirks of genetic transmission, environmental changes, chance developmental events and so forth. Finally, I show that the nature of cultural reproduction undermines the distinction between productivity and resemblance, hence it challenges the existence of forms of selection in the cultural domain.[4]

3.2 A Primer on the Price Equation

The Price Equation is widely regarded within mainstream evolutionary theory as having great power for understanding change in populations. It is especially valued for its apparent ability to attribute changes between parental and offspring generations to the influences of selection and transmission bias. Sometimes it is helpful for understanding natural populations if one deliberately makes a series of simplifying assumptions one knows to be false: perhaps that organisms are asexual, or that offspring resemble parents perfectly, or that natural selection is the only force affecting the population. One reason why the Price Equation is so valued is because the simplifications it imposes in representing biological populations are minimal (Birch 2013: 15).

For my purposes a very intuitive outline of the Price Equation will suffice that dodges complications that would be essential in a fuller treatment (what follows

---

[4] This section draws heavily on Lewens (2023).

borrows from Okasha 2006, Frank 1995 and Price 1970, 1972). Imagine two successive generations within a population. I just mentioned that the Price Equation makes few simplifying assumptions, but it does make some. In this case, assume there is no migration either into or out of the population. This means that all individuals in the offspring generation have parents from the same population's previous generation. Parents might differ in terms of how many individuals they contribute to the offspring generation. This is one way to understand parental fitness. Moreover, these facts about offspring number may – or may not – vary with facts about other parental traits. Returning to the example used in Section 1.6, if the faster-running parents are also those contributing more offspring, while the slower-running parents are those with fewer offspring, then plausibly this indicates selection on running speed. If the relationship between offspring number and running speed is completely random, then intuitively there is no selection on running speed at work. Hence, Price suggests that the intensity of selection can be represented via the statistical measure of covariance, captured by the mathematical expression:

$$cov(w, z)$$

Here w is the fitness of the members of the parental generation understood in terms of reproductive output and z is whatever trait the investigator is interested in, in this case running speed.

In a simple world where parents and their offspring resembled each other perfectly (or, in the case of sexually reproducing organisms, if offspring traits were always a precise average of parental traits) then this covariance term would suffice to calculate how average running speed changes across generations. Obviously resemblance is often imperfect. If one is trying to keep track of how a population changes, one therefore needs to know how much each parent's offspring differ from it on average; that is, one needs to know the strength, if any, of 'transmission bias'. Taking account of these differences between parents and offspring makes the mathematical treatment of transmission over generations a little more complicated. To see why, imagine a situation where slow runners have just a few offspring that resemble them very closely in terms of running speed, while fast runners have very many offspring who (for whatever reason) do not resemble them in terms of running speed at all. Under these sorts of circumstances, the fitter parents (here the faster parents) also end up exerting a particularly strong influence over the average difference between offspring and their parents. This means that a calculation of the average running speed (or whatever the trait might be that one is interested in) in the offspring generation needs to be adjusted by an average of these parent-wise changes Δz, in a way that is also weighted by parental fitness w. This overall transmission bias for the population is written thus:

$$E_w(\Delta z)$$

I can now state a standard simple formulation of the basic Price Equation for a single level of selection, which gives a calculation of the change in the average value of a trait from one generation to the next, written $\Delta \bar{z}$, and which simply adds the transmission bias term to the covariance term.

$$\Delta \bar{z} = cov(\omega, z) + E_w(\Delta z)$$

Readers might spot that this version of the equation features a curly term ω, representing *relative fitness*. This is the absolute individual fitness w from the expression above, divided by the population mean fitness $\bar{w}$.

Putting this all together, consider a biologist who is trying to understand why the average value for some trait has changed from one generation to the next. The Price Equation is widely interpreted as capturing the causal impact of selection, expressed in the term that records covariance between parental fitness and the trait of interest, as well as the distinct causal impact of transmission bias, captured via a fitness-weighted difference between values of the trait for parents and their offspring. Returning to the simple example of deer, it might be the case that fast running among parents is randomly related to their fitness. Even so, perhaps because of increasingly abundant food, it might be that offspring always grow to run faster than their parents. Here, the Price Equation delivers the intuitively correct result that change in running speed is not due to selection at all, and is entirely due to transmission bias. Meanwhile, if offspring resemble their parents perfectly with respect to running speed, then the transmission term is set to zero, and any change from one generation to the next will be attributed wholly to selection.

In the very brief exposition I just gave, I made no assumptions at all about mechanisms of inheritance. The simplifying assumptions I did rely on (about migration) were also relatively innocuous. Even so, the Equation was able to deliver sensible verdicts for the causal factors affecting a population. For these reasons, it is not surprising that the Price Equation has been regularly applied to many different contexts outside of traditional 'organic' evolution, including the domain of cultural evolution (e.g. Henrich 2004; Lehmann and Feldman 2008; Kerr and Godfrey-Smith 2009; Helanterä and Uller 2010, 2020; El Mouden et al. 2013; Birch 2017; Nettle 2020). There is now considerable – and understandable – enthusiasm for this approach, which has perhaps reached its strongest expression to date in Baravalle and Luque's assertion that, 'a certain version of the Price equation is the fundamental law of cultural evolutionary theory' (2022: 1). In keeping with the rest of this Element, while the remainder of Section 3 does not deny that the Price Equation offers a useful set of tools that can shed light on questions in the study of cultural evolution, it stops well short of giving it such a foundational status.

3.3 Price on General Selection

The idea that Price's approach to selection could be applied outside the domain of traditional evolutionary biology was pioneered by George Price himself (1970, 1995), who argued that his mathematical treatment could capture *all* selection processes, whether they occurred in nature, the marketplace, or else-where. Because of that, he gives a very diverse set of examples, intended to demonstrate just how general the recommended approach is.

One might be tempted to think that any evolutionary account of change must be focused on entities standing in reproductive relations to each other. This is encouraged by a standard practice (which I have followed until now) of discussing the Price Equation itself in terms of a 'parent' generation and an 'offspring' generation, with parents responsible for the production of differing offspring numbers. But Price thought of this as just one way in which his approach might work. He also wanted to be able to formalise the run-of-the-mill idea that an individual in a greengrocer's selects apples from a larger batch, keeping some and throwing others away. Here there is no literal 'parental' generation: the apples pre-selection do not reproduce to give rise to the apples post-selection. Instead, some apples persist and make it into the post-selection set, and others disappear. When this happens, one may find that some property of the apples – colour, or size, or something else – has changed when averaged across the population.

Price has a particular way of conceptualising this process, which fits in with his mathematical approach introduced in Section 3.2. One can think of some individual apples pre-selection as contributing, and others failing to contribute, to the post-selection population. Selection on colour occurs when an apple's pre-selection colour is non-randomly related to its contribution to the post-selection population; or, put another way, selection occurs when there is a non-random relationship between an apple's colour and its persistence. In other words, it is once again possible to represent the intensity of selection using a term $cov(w,z)$, where w now reflects the apples' persistence (rather than reproductive fitness), and z represents redness, or any other trait that one might suspect is being selected, such as ripeness, or size.

This means that Price sees selection processes everywhere: they are present in all cases where, as he puts it, 'packages' of some quantity of interest (e.g. individual apples, understood as packages of ripeness or colour) give rise to packages at a later time phase. Hence Price also points out that a series of flasks, with different concentrations of, say, saline solution, give rise to a 'selection' if different amounts are poured into beakers. The effect is that some contribute more than others (because lots of liquid is poured from some flasks, maybe none from others) to the beaker-bound 'packages' in the later population. If more is poured from the flasks with high saline concentration, and very little from the flasks with low saline concentration, then the average concentration across all packages will increase. Here 'selection' has occurred on saline concentration.

Note that the examples Price initially uses to illustrate his approach to selection all involve *stable* entities, because in these cases complications introduced by transmission bias can be ignored. A transmission bias term would need to be added if, for example, the apples selected also deteriorated quickly, or if the concentration of the solutions in beakers was chemically unstable.

3.4 Pictures at an Exhibition

One of the more unusual examples Price uses to illustrate his approach is *Pictures at an Exhibition*, a series of fifteen short pieces of music by Musorgsky. In February 1874, a memorial exhibition was held for the architect and artist Victor Hartman, who had been a friend of Musorgsky. Vladimir Stasov, a member of Musorgsky's circle who worked in the art department of the St Petersburg Public Library, and who was one of the co-organisers of the exhibition, explained that: 'Musorgsky, who loved Hartman passionately and was deeply moved by his death, planned to 'draw in music' the best pictures of his deceased friend, representing himself as he strolled through the exhibition, joyfully or sadly recalling the highly talented deceased artist' (Stasov, as quoted in Russ 1992: 16). Musorgsky's suite is structured with a repeating 'Promenade' theme, which is meant to correspond to his walk around the exhibition, interspersed among ten pieces corresponding to specific pictures from the exhibition.

Only 11 of the 400 works exhibited were chosen by Musorgsky as the basis for transformation into musical pieces. In case this sounds confusing, note that two of the paintings were jointly represented in one piece. Hence, eleven works give rise to ten pieces. Price thinks of this – reasonably enough – as a selection by Musorgsky. Recall that an apple either contributes (i.e. it is retained) or not (i.e. it is rejected) based on some property or another, such as colour. A flask contributes to some degree (i.e. its contents are transferred or not to a beaker) according to the concentration of solution in it. Price's idea is that Musorgsky selects some paintings that will give rise to music, and others (in fact, the great majority) that will not, based on some feature that makes them appealing. He suggests that: 'if one could define interesting attributes of mood and subject matter that could be quantitatively evaluated in the paintings, one could measure (using definitions of 'selection intensity' given in Price 1972) the degree to which Musorgsky selected for or against these' (Price 1995: 394).

Once again, Price's idea is that one can assign a number to each painting that represents how productive it is with respect to later music. He offers two ways of doing this: one could assign a value of one or zero according to whether a painting gives rise to a corresponding piece or not; alternatively, one could assign a number that represents how many bars of music are present in a picture's corresponding piece (which, again, will usually be zero). If one also assigns further measures to the paintings that quantify traits of interest – it could be something very simple like how much red they have in them, but also something more challenging to quantify like how sad their mood is – then by looking at covariance between productivity and redness, or between productivity and sadness, the analyst would get a sense of whether there is a selection of red paintings, or sad paintings.

The way Price uses this example highlights an important feature of his treatment of selection. In his opening example, apples in the 'parental' set give rise to apples (in fact, the very same persisting apples) in the 'offspring' set. But the elements in each set do not have to be of the same type. Pictures in the 'parental' set give rise to entirely different sorts of things – namely musical compositions – in the 'offspring' set. This means that selection, for Price, occurs even when the question of resemblance between parents and offspring makes no sense, hence even when there is no possibility of tracking how some trait of interest changes across time. Suppose, for example, I suspect that Musorgsky, an infirm man, wrote pieces for the eleven pictures that were the shortest walk from a particular spot in the original exhibition space, and ignored all the rest. Price's approach could quantify this selection on position of the original pictures, by recording covariance between distance from a particular spot and number of bars of music the painting gives rise to. However, since Musorgsky's compositions are (i) musical pieces rather than physical pictures and (ii) entities that did not even exist until after the exhibition closed, it makes no sense to attribute qualities like 'position in the exhibition room' to the musical compositions themselves.

It is worth lingering on this point, because standard versions of the Price Equation do require that the property of interest can be measured in both pre- and post-selection generations. Okasha (2006: 24), for example, follows

Rice (2004) in noting that, 'parental and offspring entities do not even have to be of the same type, so long as the character z is measurable on both'. This requirement follows from the fact that the investigator is usually interested in how some property changes in the pre- and post-selection populations, and from the consequent need to represent the effect of transmission bias. The transmission bias term relies on recording differences between parents and offspring with respect to the focal property.

Nonetheless, while the complete Price Equation does require that a property can be measured in both parent and offspring generations, Price's conceptual approach to the specific phenomenon of selection does not require this. For Price, it makes sense to understand and quantify selection in a manner that is entirely independent of how one understands and quantifies transmission bias; so much so that his approach to selection can be applied in cases where his approach to transmission bias cannot. This relies on a strict distinction between how productive an entity is with respect to the next 'generation', which is the only question that matters when determining selection, and facts about transmission bias, which only become relevant in cases where pre- and post-selection entities also resemble each other in some respect.

In the organic realm there is usually a straightforward way to apply the distinction between (i) how productive an individual is, measured in terms of offspring number and (ii) the extent to which offspring resemble their parents. Price's general approach to selection aims to maintain this key distinction between productivity and resemblance, albeit in an enlarged way. One can distinguish, for example (i) whether an apple persists or is discarded from (ii) whether the apple rots or ripens. One can distinguish between (i) how much a flask contributes to a given beaker and (ii) whether the solution contributed is stable in terms of its concentration. And one can distinguish between (i) how many bars of music a painting yields and (ii) whether the music in question is even capable of instantiating important properties of the painting.

3.5 The Price Equation in Cultural Evolutionary Theory

Some ways of applying the Price Equation in the context of cultural evolution are similarly straightforward. That is because they focus on the extent to which what an individual has learnt covaries with how many *biological* offspring it has. This is what Jonathan Birch (2017: 197) calls type-1 cultural selection, or CS_1. The approach to an entity's productivity is the same here as that used in the standard biological context, hence it can also be applied independently of questions about resemblance between parents and offspring.

As I explained in Section 1.6, Birch's CS_1 is just one approach to understanding cultural selection. In other cases, instead of focusing on how learned traits affect an individual's production of biological offspring, investigators aim to understand an individual's production of 'cultural' offspring. Birch calls these latter approaches type-2 cultural selection, or CS_2 (2017: 199). The underlying motivation for CS_2 can seem intuitive enough: even if they have no biological offspring, some individuals nonetheless have plenty of cultural offspring, in the sense that their influence on the cultural traits of subsequent generations is significant. But remember, once again, that the sort of approach presupposed by the Price Equation demands that two questions can be distinguished (Okasha and Otsuka 2020): (i) how many offspring (cultural or otherwise) does some entity have, and (ii) to what extent do the offspring resemble their parents? This distinction is far less clear-cut than one might think in the domain of culture, especially in the context of CS_2 approaches. This is because of features of cultural 'reproduction' stressed by theorists of cultural attraction (Sperber 1996, 2000). Indeed, the very idea of determining cultural 'parents' entirely independently of cultural resemblance is challenging. This, in turn, illustrates problems with the notion of cultural 'fitness', and hence with the idea of cultural selection.

El Mouden et al. (2013) have developed an approach to using the Price Equation in the context of CS_2. Their formulation of the Price Equation for culture mirrors the standard formulation for organic evolution exactly: indeed, the equation is written in the same way in both cases, albeit with the letter c (rather than ω) representing some measure of cultural 'fitness' (2013: 233):

$$\Delta \bar{z} = cov(c, z) + E_c(\Delta z)$$

The intuitive idea behind this is, again, simple. Suppose a population is once more divided into parental and offspring 'generations'. These needn't be biological parents (any more than Price's pre-selection apples are parents of the post-selection apples), or indeed 'parents' in any standard sense of the term. Instead, each individual is thought of as productive of some number of cultural descendants: some individuals may have none, others many. Moreover, the transmission bias term is also required to capture differences that arise when descendants resemble their cultural parents imperfectly.

At first sight, this approach might seem to work without too much trouble. Perhaps some individuals in a population – think of them as social influencers – have lots of disciples, and those disciples aim to mirror the hairstyles of their heroes. Other individuals have next to no one paying them any attention, and no one makes much effort to mirror their hairstyles. One can then see how hair length changes from one cultural 'generation' to the next: cultural selection on hair length will be very strong if individuals with lots of disciples also have longer hair, while individuals with few disciples have shorter hair. Cultural selection might be overwhelmed if, for some reason, the efforts of disciples to mirror the hairstyles of their heroes frequently fail. Maybe they attempt to style their hair at home, and they end up cutting far too much off. Under these circumstances the term that represents 'transmission bias' will be high, because offspring tend not to resemble their cultural 'parents' very closely.

3.6 Cultural Productivity and Cultural Resemblance

In spite of the appeal of simple examples such as this one, applications of the Cultural Price Equation face a dilemma when it comes to determining cultural fitness. It is often the case that one can only give a plausible answer to questions about who a given individual's cultural offspring are if one also makes use of information about cultural resemblance; however, if selection is interpreted in the very pure way that the Price formalism demands then these two domains are supposed to be kept entirely separate.

To see this, imagine that Shy Simon invents a wonderful gadget with features ACE. Meanwhile, Outgoing Oswald invents a mediocre gadget with features URG. Suppose, further, that no one pays attention to Shy Simon, but lots of people pay attention to Outgoing Oswald. In spite of paying lots of attention to Oswald, many folk end up constructing a gadget just like Simon's, with features ACE. No one constructs anything with features URG.

This could happen because a gadget with features ACE has the following properties: it answers a widely experienced need; it makes use of elementary and intuitive design principles; and it can be built easily using cheap and plentiful materials. It is, in Sperber's sense, an 'attractor'. Meanwhile, a gadget with features URG has all the opposite features: no one has much need for such an item, its mode of operation is not at all intuitive, its raw materials are expensive. Even so, it's not out of the question that individuals observe Outgoing Oswald, and they think 'I can do so much better than his URG!' It is because they observe Oswald that they end up building gadgets with features ACE.

I do not think this kind of case is so far-fetched that one should dismiss it as irrelevant: after all, the Price approach is supposed to be so general as to cover all instances of selection. The case raises problems for efforts to apply this approach – and more generally for notions of cultural fitness and selection – to culture. First, it is not the case that the individuals who produce gadgets with features ACE are Shy Simon's cultural *offspring*, even though their gadgets resemble his closely. Shy Simon's gadget plays no role in explaining the genesis of the resembling token gadgets. One could, therefore, try to argue in ways that parallel what might be said about organic reproduction above: an individual can have plenty of babies which, for whatever reason, fail to resemble their parents. The Price approach would then give a peculiar and unilluminating result, namely that this is a case with intense selection of features URG (which by hypothesis no one is impressed by) accompanied by very strong transmission bias. While Oswald's gadget – unlike Simon's – has a causal role in explaining the production of these later tokens, that role is small. There are many other factors that also explain (causally) the production of later tokens – everything from the availability of raw materials to the intuitive nature of the design. Crediting Oswald with high cultural 'fitness' in a case like this seems to give him a spuriously inflated role in explaining the widespread adoption of ACE.

3.7 Cultural 'Influence'

I have argued that Price's general approach to selection relies on a strict distinction between the question of how productive a pre-selection entity is with respect to the post-selection set, and the question of whether elements of the post-selection set resemble elements of the pre-selection set. In Price's example of Musorgsky, some of Hartmann's pictures are productive with respect to the generation of music, others are not; but these facts are supposed to be determinable independently of any question of whether the music resembles the pictures.

El Mouden et al. abandon this strict distinction. Rather than keeping the questions of whether an individual has cultural offspring distinct from questions of resemblance, they are instead blurred. They begin by defining cultural ancestry as follows: 'Person A is a cultural ancestor of person B if the value of z person B has was influenced by the value of z person A had' (2013: 233). This definition of ancestry in terms of influence departs from Price's treatment. The same may be true of the approach suggested by Kerr and Godfrey-Smith (2009: 533). They rightly note that the Price Equation requires that lines of 'connection' can be drawn between parents and offspring. In the biological context these are usually reproductive relations. Kerr and Godfrey-Smith continue: 'we have discussed connections mostly as parent–offspring relations ... Alternatively, a connection may represent other forms of *influence* between entities such as material or information flow' (emphasis added). But this notion of connection as 'influence' once again threatens to undermine the distinction that is central to Price's approach, because an intuitive way to understand 'influence' is partly via the notion of resemblance of token entities across generations. To see this, recall the earlier discussion of biological cases where inheritance is exceptionally unreliable: fast runners have lots of offspring, slow runners have very few, but fast running parents fail to have fast running offspring, and slow running parents fail to have slow running offspring. The investigator does not determine which organisms are an individual's parents by asking which organisms have influence on the individual's running speed: indeed, this is a central case of selection for running speed being strong even though individuals are not 'influenced' by their parents at all in this respect.

El Mouden et al. move on to suggest that 'cultural fitness is a measure of cultural influence, reflecting both the number of people who learn from an individual, and the degree to which their traits are influenced when they do learn' (2013: 233). Again, this idea of equating cultural fitness with cultural influence on traits of the offspring generation goes against Price's background conceptualisation of selection: for here, recall, a painting might have high cultural fitness merely because it gives rise to many bars of music. The question of whether traits of the music are 'influenced' by the picture does not come into it.

El Mouden et al. say a little about how they understand this notion of cultural influence – specifically, they claim that an individual may be influenced by many other individuals, to different degrees, for a specific trait – but this does not go far to explaining what counts as strong or weak influence. Suppose, for example, that I am so disgusted by an individual's behaviour that I try my best to act in the exact opposite way: does the individual in question influence me strongly, because their effect on my life is marked, or weakly, because I do not aim to be anything like them? One might argue that if an individual is highly influential with respect to future generations, this means that because of exposure to that individual, those in subsequent generations act or look a certain way that approximates to that of the influencer. Again, this would defy what the Price Equation asks us to do, which is to distinguish between (i) the productivity of an entity strictly with respect to how many offspring it generates, whether an entity persists or not, whether it gives rise to bars of music, and so forth and (ii) the degree to which the post-selection entities correspond to the pre-selection entities.

3.8 Causal Agnosticism

Tim Ingold (2022) has recently argued for the elimination of the notions of cultural transmission and cultural inheritance. He holds that these notions, allied to related concepts of genetic inheritance and genetic transmission, give the impression that learning is a simple matter of passing behaviours from one cultural generation to the next, in the same way that one individual might 'transmit' a message to another by handing them a letter, or in the way one individual might 'inherit' a house, or a collection of stamps, fully intact from another. In describing cases of learning using these terms, one diverts attention away from the ongoing processes by which agents actively recreate the domains into which they are born.

I don't deny that notions like cultural transmission, and cultural inheritance, may have these unfortunate connotations. But within the Price framework, the problem is not that terms like 'transmission' and 'inheritance' give a misleadingly passive impression of the processes underlying cultural change and stasis. The problem is rather the opposite one, namely that 'transmission' and even 'inheritance' are entirely agnostic when it comes to the nature of these causal processes. 'Transmission bias' occurs when learning processes (regardless of how active, interpretative, collaborative and reconstructive they may be) give rise to dispositions in learners that depart in some way from their models. 'Transmission' is free of bias, alternatively 'inheritance' is faithful, if precisely the same processes – again, no matter how active or inactive they are – result in resemblance.

A contrived example will illustrate the difficulties this causal agnosticism poses.

> **Bake Off!** Baking 'influencers', with many disciples, manage to invent a new baking technique YUM. YUM is responsible for producing light, fluffy cakes. Meanwhile, individuals with only a tiny handful of disciples invent a baking technique YUK. Use of YUK yields heavy, cloying cakes. The different bakers' disciples all manage to re-produce precisely the techniques of their mentors, and average lightness of cakes increases in the population.

There are many different possibilities for why the disciples' cakes resemble those of the influencers, and the Price Equation does not discriminate between them. This is a limitation of the approach. Here are just two such possibilities that might explain why resemblance occurs:

1) **Slavish copying.** All the disciples set out with the goal of doing precisely what their heroes do. They study their heroes' YouTube videos and copy their every movement slavishly.

2) **Loose inspiration.** Disciples all admire their respective baking heroes, but they make no effort to copy their techniques bit-by-bit. Even so, it turns out that disciples of the bakers who use YUM all end up formulating YUM, and disciples of the bakers who use YUK all end up formulating YUK. YUM and YUK are both attractors: they are easy to devise and enact, and they both have superficially tempting features. Resemblance occurs because the majority of bakers in the offspring generation are highly talented at baking,

a few are not so good. The more talented bakers – because of their good judgement – also pick their heroes well. They admire bakers who are talented enough to invent YUM, while poor bakers end up admiring those who instead are only good enough to invent YUK. And because of this difference in skill, when the better bakers in the offspring generation set out to devise a technique, the result of their talents is that they also develop YUM, while the poorer bakers develop YUK.

El Mouden et al.'s approach demands that both scenarios be analysed in exactly the same way. Because offspring resemble their cultural parents perfectly with respect to their baking techniques, the transmission bias term must be set at zero regardless of what explains why resemblance is so close. The Cultural Price Equation therefore mandates that in both cases all the population change is attributed to cultural 'influence' from the parental population, which (for them) is synonymous with cultural selection.

This is a shortcoming: it seems important to find some way of recognising important differences between the two scenarios. First, the influence of the parental population is lower in scenario two than scenario one, because in the 'Loose Inspiration' case the offspring individuals' own creativity – rather than the influence of their heroes – takes more responsibility for the change in the direction of lighter cakes. Second, it seems a mistake to attribute *all* of the overall population change to selection in this second scenario. What happens here is talented bakers (i) admire other talented bakers and also (ii) they are better at discovering valuable baking innovations; conversely, untalented bakers admire other untalented bakers, and they are worse at discovering valuable innovations. This looks like what one might call a 'false positive' for Price's approach: it meets the criteria for there being selection, but on inspection it is not clear that this really counts as a case of selection at all.

This example shows that nuance must be added to El Mouden et al.'s claim that, ' . . . an important part of cultural evolution is the transmission component —which reflects the action of minds that have been shaped by natural selection to process information in ways that enhance genetic fitness' (2013: 237). The transmission component may indeed reflect the action of minds in this way, but the selection component – in Price's sense of 'selection' – can also reflect the action of minds in the very same way. That is because the Price approach understands 'selection' to have occurred merely when offspring traits resemble parental traits, thereby making the transmission bias term small. The creative use of individuals' minds can potentially bring about both divergence and convergence when those individuals approach the same problem as their cultural 'parents'. Divergence is likely when there are many viable solutions, the problem itself is only vaguely specified, and there are few constraints on the approaches likely to be followed. Convergence is more likely when there are only a few viable solutions, the problem itself is tightly specified, and approaches likely to be followed are highly constrained.

El Mouden et al. go on to say that, 'the disagreement between those who advocate a Darwinian or a non-Darwinian approach to cultural evolution comes down in large part to different views about the relative importance of selection versus transmission in cultural change' (2013: 238). Again, this isn't quite right: at least some disputes – for example, over the relative importance of cultural 'attraction' – do not concern how important the transmission term is. In the 'loose inspiration' variant as laid out above, one reason why transmission bias is so low is because good bakers find it easy to recreate YUM when they have been inspired to do so by their heroes. The ease with which it is recreated may be because YUM is an attractor.

It is now easier to understand why some advocates of cultural attraction theory complain that it is misleading to equate 'attraction' with 'transmission bias' (as El Mouden et al. seem to do). Some advocates of cultural 'attraction' instead point out that (as I have just explained) various forms of creative inference can potentially underpin cultural change whether the transmission term is significant or not (Scott-Philipps et al. 2018). The worry here is not that selection might be less important than transmission; it is that the distinction between selection and transmission – which the Price Equation encourages the analyst to understand as distinct causal factors represented by the two terms on the right-hand side of the equation – is a misleading one because it does not map neatly onto underlying cognitive processes. If prolific cultural parents have traits that correspond to attractors, then attraction itself can reduce transmission bias and underpin faithful reproduction between parents and offspring. On the other hand, if prolific parents have traits that do not correspond to attractors, then these factors of attraction can account for systematic differences between parents and offspring. In this way, the very same factors of cultural 'attraction' can underpin both cultural selection and transmission bias.

3.9 Diagnosis

The Price approach is susceptible to these difficulties because of an important way in which cultural 'reproduction' differs from organic 'reproduction' (Scott-Phillips et al. 2018; see also Nettle 2020 for some sceptical caveats). Price's basic approach, as I have stressed, relies on a strong distinction at the level of underlying processes between (i) the question of how many elements in the post-selection set are produced by elements in the pre-selection set and (ii) the question of the extent to which elements in the post-selection set resemble elements in the pre-selection set. In the organic context, this distinction is usually unproblematic. For example, on the productivity side one might judge that organism A has four offspring, while organism B has just two; meanwhile, one might also judge that organism A's offspring resemble organism B, and vice versa.

In Price's motivating examples of non-organic selection among apples, or flasks of liquid, the same distinction is also unproblematic: it is perfectly coherent to judge that apple A persists, while rotting in such a way that it ends up with the qualities of texture and taste had by apple B. In such cases there are easily trackable entities (a persisting apple, a corresponding beaker into which a flask is poured) that allow the analyst to distinguish in a neat way between how 'productive' an entity is, and how much its products resemble it. The same distinction is far harder to apply in cultural contexts, where the very idea of cultural 'offspring' is hard to untangle from some notion of cultural resemblance. For suppose an individual A produces a token behaviour that shocks many other individuals; they try to overtly display something very different. Should one say that although A has many cultural offspring, those offspring end up resembling a wholly different agent? If one does say this, absurdities follow. The whole point is that while observers are certainly affected by A, there is no helpful sense in which A is being culturally selected, or has high cultural 'fitness'. Observers are deliberately avoiding any repetition of A's behaviour.

The alternative is also unpalatable. One can assign a low cultural fitness to A, on the grounds that behaviour resembling A's does not appear reliably in the subsequent cultural generation. This leads to a different problem: one is no longer drawing the sort of distinction the Price approach mandates between selection and transmission bias. Instead, the assessment of A as having low fitness reflects an amalgam of both the amount of 'productivity' A has, in the sense of the number of people who are in some way affected by A, and also the fact that A's cultural 'offspring' are usually nothing much like A. In other words, the Price approach either gives a highly misleading result in terms of understanding change in the population, or it gives the less misleading result at the expense of making use of a notion of cultural fitness (and thereby of cultural selection) that is a distortion of the way that notion is deployed in mainstream evolutionary theorising.

The argument of this section does not show that all invocations of the Cultural Price Equation are misguided. Instead, it gives theorists who wish to use it two options, both of which complement the flexible outlook on cultural selection defended in Sections 1 and 2. One is to acknowledge that the very ideas of cultural selection and cultural fitness need to be applied in loose ways that sometimes depart considerably from those endorsed by this formal approach. The other is to ensure that if they are to be applied more strictly, this happens only in those domains of cultural evolution (such as Birch's CS_1) where one can draw clear distinctions between facts about productivity and facts about resemblance.

4 Waiting for Casabe

4.1 Manioc Processing in Amazonia

The previous sections each began by addressing a very general question: how to define cultural selection, how it might relate to cultural attraction, and the prospects of formalising it via the Price Equation. This final section instead begins with the puzzles posed by a specific case, namely the processing of manioc in Amazonia. My primary goal here is not to show how the general lessons of the previous sections should be put into practice. Instead, my aim is to show how the strategy of beginning with a specific case can generate further lessons, which complement those of the previous sections, for a general approach to cultural selection.

Manioc – also known as cassava – is an important food in many parts of the world. Its leaves can be eaten; however, it is the starchy roots that constitute its most important dietary element. They serve as a staple for at least 800 million people worldwide (Food and Agricultural Organization of the United Nations 2013). All of manioc's many varieties contain – in both roots and leaves – 'cyanogens'; that is, chemicals with the potential to release highly toxic hydrogen cyanide. The 'sweet' varieties of manioc contain low enough levels of cyanogens in their roots for simple scraping and boiling to render them safe to eat. This is not, however, the case for the 'bitter' varieties, which have much higher cyanogen concentrations. Without proper processing, their consumption leads to a range of serious diseases.

Many cultures have devised elaborate multi-stage processing regimes – involving various combinations of scraping, grating, cooking, soaking, washing and waiting, often spread over many days – that succeed to greater or lesser extents in removing manioc's toxic cyanogens (Dufour 2006). The specific approach to processing taken by Tukanoan people in Northwest Amazonia has become iconic within cultural evolutionary circles, thanks to the use Joseph Henrich (2016) makes of their case. The manioc they prefer has levels of cyanogens in its roots that are high even by the standard of the 'bitter' poisonous varieties. And yet, there is no evidence that they suffer from either chronic or acute health problems linked to cyanide poisoning (Dufour 1994). In other words, Tukanoan people have set themselves an especially challenging nutritional task, which they have also solved.

The Tukanoan case is particularly interesting for the way in which it appears to show cultural selection acting as a 'hidden hand'. Consider first how natural selection – rather than cultural selection – is often presented. One might wonder why bitter manioc tubers contain such high levels of cyanogens. The answer seems to be that these chemicals – which are only released when cell walls are damaged – are an evolved defence against many destructive species ranging from beetles to rodents (Wilson and Dufour 2002). Darwin's (1859) mechanism of natural selection shows how chance variations can proliferate in a population, and serve as bases for further improvement, when they contribute to the fitness of the organism. Natural selection works as a hidden hand because no overseeing agent need control, register or understand the nature of the benefits conferred by variations that arise through chance alone. Manioc roots do not know that their cyanogens have the capacity to poison destructive creatures. Darwin even goes so far as to argue that the relentless scrutiny of natural selection makes it a far superior process of adaptive change compared with the intelligent use of artificial selection by plant and animal breeders.

To understand why the Tukanoan case has become iconic, and why it might appear to show cultural evolution acting as a hidden hand, it is important to have a picture of how they – or at least the group studied by Darna Dufour, on whose original fieldwork Henrich relies – process manioc. They use the roots for various purposes, including making a kind of bread called 'casabe'. Having dug the tubers from the ground, they remove the outermost layer of peel, then wash and grate the roots. This produces a sort of watery mash. The mash is washed and squeezed in a strainer, which removes the fibre from the starch and liquids. The starch is then allowed to settle from the washing water. The latter is boiled straight away to make a drink called 'manicuera'. The starch and fibre, however, are stored at least overnight, but ideally for 48 hours. They are then recombined and baked to produce casabe.

Experiments have shown that casabe baked on day three (i.e. when the fibre and starch have been left for the full 48 hours) contains only 3.2 per cent of the cyanogens contained in the whole unprocessed roots. If one tries to shorten the process, leaving the fibre and starch for just 24 hours, then this percentage is over twice as high at 7.4 per cent. And if one doesn't wait at all, then the percentage is 7.6 per cent. In other words, patience pays off in terms of detoxification (Dufour 1994, 1995).

Henrich suggests that there is no easily detectable signal of the efficacy of this final drawn-out waiting stage. In his telling, there is a good perceptible proxy for cyanogen content in the early phases of processing, because the manioc becomes less bitter. But the two days of waiting are not accompanied by any diminution in bitterness. Insightful individuals might therefore be tempted to skip this final waiting stage and get the chance to eat their casabe faster. Not only would the product taste good, the poisonous nature of what one might call 'fast' casabe would also be exceptionally hard to detect, because chronic cyanide poisoning takes such a long time to manifest itself. Acute cyanide poisoning has immediate effects, but levels of cyanide are low enough in fast casabe that only the longer-term chronic effects are in play. No one would be able to link the increasing numbers of ill people to the way they have been making their bread.

These features have made Henrich's use of the manioc case into an important exemplar for how to think about cultural accumulation (Mercier and Morin 2019). Sterelny, for example, says that Henrich 'has pointed out that social learning is very important when the challenge is causally opaque. His most compelling example is manioc processing' (2021: 39). I, too, have noted how Henrich makes the case that 'a form of "hidden hand explanation" modelled on organic selection' must be invoked to explain manioc processing (Lewens and Buskell 2023). Just as Darwin had previously argued for the superiority of natural selection to the eye of the intelligent breeder, likewise it seems that cultural evolution can be superior to intelligent individual insight.

4.2 Cumulative Cultural Evolution

Like Richerson and Boyd before him (see Section 1.1), Henrich uses historical examples of European explorers who become isolated or lost in unfamiliar environments to motivate his account of cultural accumulation. These explorers suffer miserably – often they die – because of a lack of know-how. If they are wise, they throw themselves on the mercy of people whose local knowledge has equipped them to find and process nutritious food, to construct shelters, to avoid dangerous creatures and dangerous terrain, and so forth.

These examples draw his readers' attention to a series of important themes that I also stressed in Section 1.1. First, humans cannot rely on instinctive knowledge to tell them which foods are safe to eat, how to protect themselves from the elements, and so forth. These things must be learned. Second, even the most insightful individuals are unable – perhaps through lack of time, perhaps through lack of sufficient lucky insights, perhaps through lack of background knowledge – to invent a full suite of survival techniques that will serve them in unfamiliar environments. When they learn what is needed for survival, they must learn it from others. Third, these themes apply to local initiates no less than they do to unfortunate interlopers. Individuals perpetuate a set of practices that has accumulated over time thanks to the insights and good fortune of others before them. In these senses, survival relies on culturally accumulated wisdom.

I endorse all three of these lessons, hence I endorse the claim that Tukanoan manioc processing is an instance of cumulative cultural evolution. I do, however, suggest some important supplements to Henrich's story. First, I point to a problem that arises if too much stress is placed on causal opacity. It makes the initial generation of effective cultural traditions mysterious. Second, while there is evidence that manioc processing is learned by copying the previous generation, I argue that this copying should not be described as wholly 'dumb'. Third, I further cement the effort of this Element to draw links between cumulative cultural evolution and the theory of cultural attraction. The manioc case is an instance of cumulative cultural change. But it is also a case suggestive of individual insights, and it indicates the likely ways in which very widely held preferences for fermented foods – that extend beyond the bounds even of our species – may have played an important role in explaining the uptake and maintenance of the final steps of the manioc processing package.

4.3 Ignorance That and Ignorance How

In Chapter 5 of *The Secret of our Success*, Henrich (2016) sketches a useful 'toy' example that illustrates some general principles for how cultural accumulation works. It shows how community-wide processes of observing and learning from others can give rise to suites of adaptive techniques that no one individual is ever likely to invent on their own. In Section 4.4, I will show that there is a tension between this toy model, and the stronger claims made for causal opacity in Chapter 7 of Henrich's book.

Henrich imagines a group of organisms – in his telling they are primates – that need to extract food from their savannah environment. Whether by its own limited inventiveness, or just by luck, one individual starts using a stick to get nutritious termites from a termite mound. Others see how well this technique works, and they copy it; but among those copiers, one mistakenly thinks the stick has been sharpened, and tries to sharpen their stick, too. The sharp stick works even better, and so later learners tend to copy this. As the story goes on, additional elements are added to this suite of behaviours. For example, someone plunges a sharp stick into a termite mound as usual, and ends up spearing a rodent. Meanwhile, someone else notices they can find rabbits by following their tracks. Yet another individual, in a later generation, notices the effects of both of these behaviours, and combines rabbit-tracking with rodent-spearing. As a result, they can hunt rodents down in their burrows. The story continues, with a final act in which later individuals copy a whole set of their predecessors' efficacious savannah hunting techniques, none of which the latecomers had any role in inventing. As Henrich summarises, 'this is a toy example meant to illustrate how selective cultural learning can generate a cumulative evolutionary process that generates cultural packages that are smarter than their bearers' (2016: 55). Note, once again, that Henrich does not refer to 'cultural selection' here for reasons explained in Sections 1.6 and 1.7. This explanation is, however, quasi-selectionist in the sense that it illustrates how the beneficial variation that arises in a population can be preserved and built upon when individuals in later generations selectively copy their predecessors. This is the same general lesson that was exhibited by the example of scientific research and education that appeared back in Section 1.1. Learning from others allows the many small discoveries that can be attributed to individuals (and increasingly to coordinated and organised groups of individuals (Derex 2022)) to be retained, pooled, absorbed and then put to work by further individuals who may have little or no idea of the forms of work, reasoning or just plain luck that went into those discoveries in the first place.

There is plenty of room for individuals to have modest insights in Henrich's toy example. As he tells the story, one individual might figure out that a stick can be sharpened, and persists with this technique because they see it works better that way. Another individual recognises this success, and uses the same sharpened-stick technique because of that. Further down the line, individuals chose to adopt whole suites of techniques that they see being used successfully by others. In other words, agents are well-placed to tell *that* some technique works well, even if they can't say *why*. This is not a feature restricted to 'folk' or 'lay' epistemologies: a large-scale clinical trial can also determine whether a new drug is efficacious, even though pharmacologists may remain ignorant of the mechanism that explains the drug's potency.

By stressing the difficulty of detecting the long-term poisonous effect of 'fast' casabe, Henrich takes the significant further step of arguing that it is extremely difficult for anyone to figure out even *that* the 48-hour wait has a detoxifying effect, or indeed any valuable effect at all, let alone *how* it has that effect. So this case appears to demonstrate in a particularly acute way that cultural evolution is 'smarter' than any individual. It can construct techniques that are highly efficacious, even in cases where no one has been aware that this is the case. In this respect, it is like natural selection, which can make manioc roots repel destructive pests without anyone ever needing to know that roots have this effect.

Although this way of telling the story brings cultural evolution closely into line with natural selection, it also makes it hard to see why the various stages of manioc processing became adopted in the first place. Henrich's toy example allows that both the original innovator, and those who copy what they do, are in a position to notice that the stick poked into a termite mound has a positive impact on the termites gathered. In cases where causal opacity is so extreme that neither the originator nor the copier is aware that the technique works, one needs to explain why the originator persists with the technique, and why others think to try it as well.

4.4 Learning from One's Predecessors

Henrich's primary purpose in calling on the manioc case is to argue for the importance of copying the wisdom of previous generations in an uncritical or unreflective way that makes cultural evolution, like natural selection, 'dumb' (2016: 12). If individuals are unable to figure out that some stages of processing are effective, then it is pointless for them to attempt a survey of their companions, in the hope of copying the specific elements of processing behaviour they judge to work well. Instead, they are better off attending to individuals who are likely to have already solved the processing problem, with the goal of copying them wholesale.

Consistent with this account, Tukanoan women often have little to say about many of their choices, including why they cultivate such toxic varieties of manioc in the first place. As Dufour puts it, 'Tukanoan women each explained it in terms of tradition: their mother did it that way, and their mother's mother before her' (2006: 7). This deference to tradition offers a potential way to answer the puzzle posed in Section 4.3. It brings cultural evolution even more closely into line with natural selection, by drawing on reliable (albeit uncomprehending) inheritance to assemble and preserve adaptive traits over time. Perhaps Tukanoan women repeat an action pattern whose efficacy they are unaware of, simply because they are motivated to do whatever their mothers did before them. This proposal raises two questions. First, how can imitation of what goes before solve the problem of the origination, rather than merely the maintenance, of efficacious techniques? Second, to what extent is this imitation 'dumb'?

4.5 Cultural Ratcheting

If cultural repertoires are to accumulate over time, they must show what theorists working in this area call a 'ratchet' effect (Tomasello et al. 1993; Tomasello 1999; Tennie et al. 2009). This idea of cultural ratcheting is just a redescription of the basic idea of cultural accumulation; namely, that what is gained in one generation is not lost in the subsequent generation, but instead it is built on and elaborated. It is trivial that learning under the influence of others must have some role in this process (Mesoudi and Thornton 2018; see also Buskell 2022). What is not trivial is to spell out exactly how this ratcheting occurs.

As noted in Section 2.5, reproduction of cultural repertoires at the level of the community in question must be achieved somehow if accumulation is to occur (Lewis and Laland 2012). There are several potential explanations for how this might happen. It might be that each individual learns each technique through faithful observation of another individual, making community-level reproduction a consequence of many instances of individual-level reproduction. It could also be that forms of 'attraction' stabilise community-level traits, when each individual learns from repeated exposures to a pool of other individuals. It could be that individuals in one generation collectively maintain – and thereby curate, whether knowingly or not – an environment that canalises the learning of the subsequent generation. And, of course, complex hybrids of all these modalities – and of others not listed here – are also possible.

Research has supported an eclectic range of answers to these questions. For example, Enquist et al. (2010) have argued for the necessity of learning from several different models (cultural 'parents' in their terminology) if community-level inheritance is to be assured. Caldwell and Millen (2009) have suggested that detailed attention to the precise actions of models is not necessary for cumulative cultural evolution. Sterelny (2012) has stressed the importance of what he calls 'apprentice learning', whereby neophytes have augmented opportunities to learn because they find themselves in environments that are felicitously structured thanks to the tools, prototypes, discarded by-products and so forth left around by adepts.

A full review of all this work is well outside the scope of this Element. Instead, I want to look specifically at what Henrich has to say about the manioc case. He suggests that 'over-imitation' is one of the significant keys to explaining how cultural ratcheting occurs. This is a technical term that describes the faithful copying of a full series of actions that precede the attainment of some goal, even when the contribution of some of those actions to the goal may be completely redundant (see Hoehl et al. 2019 for a survey). In a classic experiment, children were asked to do whatever was required to get a reward – it was a sticker – from a Perspex box. A demonstrator had shown them a way of doing this, which incorporated several superfluous actions. The experiment was designed to make the redundancy of these actions obvious to the children. Even so, they tended to copy all the demonstrator's actions, when they could have obtained the sticker more directly (Horner and Whiten 2005).

Subsequent experiments have shown that this 'over-imitation' is not limited to children, and can be even more pronounced among adults (Hoehl et al. 2019). In Horner and Whiten's original experiment, chimps did not imitate actions they saw to be causally irrelevant: they just got their reward in a direct way. This has given rise to a widely accepted story that gives the individual stupidity of humans compared with chimps a key role in explaining why we are such effective cultural innovators compared with them (Boyd and Richerson 1996). Henrich's thought, following this work, is that Tukanoan women copy what their mothers do, even though the final waiting steps have no discernible relevance to how casabe turns out.

Although Dufour records Tukanoan women explaining that they simply do what their mothers did, this does not mean that they copy their predecessors in a wholly blind way. It is not possible to reproduce literally every aspect of another's behaviour to an arbitrarily fine level of detail. As Charbonneau and colleagues have pointed out in a series of papers, the description of learning methods is subject to a 'grain' problem (Charbonneau 2020; Charbonneau and Bourrat 2021). Suppose my mother tells me a story – the story of *Goldilocks and the Three Bears*, say – and I repeat it to one of my own children. In writing that I 'repeat' her telling of the story, I leave open how fine-grained my repetition is. Do I keep the same ordering and specification of major episodes, such as numbers of bears and rough temperatures of porridge? Do I go further, and repeat all the same words that she uses? Do I maintain the same intonation, and hand-gestures? Do I sit in the same way, and drink a cup of the same type of tea during the telling?

These questions have significance for a variety of issues in cultural evolutionary theory. For example, they challenge a distinction that is frequently put forward in literature on cultural evolution between 'imitation' – where one copies another's bodily movements – and 'emulation' – where one merely copies the end-point of an action. Suppose that Calvin creeps circuitously to the cupboard for a cup of cocoa. Casper copies Calvin, and Camilla copies Calvin. When Casper copies Calvin, he also creeps circuitously to the cupboard for cocoa. When Camilla copies Calvin, she collects cocoa from the cupboard, but she declines to creep and instead she canters confidently. It is tempting to say that while the first case is an instance of imitation, the second case is an instance of emulation. Camilla simply obtains what Calvin obtains. She cuts to the chase, so to speak, getting cocoa but using different bodily motions. Meanwhile, Casper reproduces Calvin's specific bodily movements.

Unfortunately, this verdict does not rest on a deep difference between the two instances of copying; instead, it rests on the level of detail the analyst packs into their description of the events. Both could be called 'imitation', if we think of copying bodily actions in very broad-grained terms. Both Camilla and Casper convey themselves to the cupboard, perhaps choosing roughly the same route; they simply differ in how they hold themselves as they do so. Meanwhile, neither will be called imitation if we think that copying bodily actions requires that the precise nuance of Calvin's creeping, the specific way he reaches for the cocoa, maybe even the precise same circuitous route, are reproduced.

Returning to the manioc case, what does it mean to say that people do what their mothers do? First, an individual's mother is unlikely to go through the exact same bodily motions each time she makes manioc; grating might be faster or slower, she might face one direction one day, another direction on another. Second, there are bound to be some aspects of the mother's performance that are not copied, even if they are regularly repeated by her. In both cases, an implicit evaluation of what to copy seems to be guided by some intuitive sense of what might, and what might not, plausibly be causally relevant: 'It probably doesn't matter that I have my left foot slightly in front of my right while grating the manioc, or that I wear a t-shirt, even though that's not what my mother usually does; it probably doesn't matter if I start the storage process a bit later in the day than she does; but it might matter that I store the manioc for roughly the same time as she does, even if I can't really think exactly what that might do.'

More generally, while the literature on overimitation supports the claim that individuals often repeat unnecessary actions – unnecessary in the sense that they don't make a causal contribution to the final effect of the action sequence – it does not follow that individuals have a blanket tendency to reproduce everything the demonstrator does, regardless of context (Hoehl et al. 2019). Visiting an unfamiliar religious site while on holiday, I might pay close attention to how everyone else there behaves, even though I suspect many of them have come to see the same artwork that I have come to see. Perhaps I copy them by making sure that my clothing is sober, that I walk slowly, that I don't speak loudly. I could just cut to the chase, and dash to the specific artwork that motivated me to come, but I imitate a much broader series of preceding actions because I do not want to look ignorant or disrespectful. Situations regarded by children as 'playful' also seem to elicit higher levels of overimitation than situations regarded as focusing on pure efficiency, perhaps for the simple reason that games often have rules that everyone needs to stick to. So the same action sequence (turning around and touching the ground, say) might be copied in the context of a game; but not copied in the context of a more utilitarian activity. Children invited to participate in psychological experiments on overimitation might want to demonstrate that they have been paying close attention to the performances they are asked to observe. And so they copy irrelevant actions such as tapping a stick three times before using it to open a box, but they do not take their sweaters off if the demonstrator is only wearing a shirt. This is not construed as a relevant part of the action sequence in the first place. In stressing the flexibility of overimitation, I do not mean to imply that its deployment is always guided by conscious deliberation, as opposed to other sources of plastic response such as associative learning. But while overimitation is real, it is not a blanket copying response to all of the actions of others (Hoehl et al. 2019).

4.6 Success, Prestige and Health

In pointing to overimitation as the way to preserve effective techniques, one defers the question of how this wisdom was accumulated in the first place. Henrich suggests how this might have been achieved at the end of his discussion of Tukanoan people: 'Operating over generations as individuals unconsciously attend to and learn from more successful, prestigious, and healthier members of their communities, this evolutionary process generates cultural adaptations' (2016: 99).

Henrich mentions success, prestige and health: I will tackle each in turn. In suggesting that Tukanoan women attend to the successful, he does not mean that women attend to those whom they see having greater success in eliminating cyanogens. The causal opacity claim implies that no one can determine who is successful in this respect, at least not if one is focusing on the reduction effected by the 48-hour waiting phase. This leaves open the obvious thought that Tukanoan women gradually change their approach to bread-making by attending to the techniques of those whose bread they most like the taste of. In fact, women do have the chance to try each other's casabe at communal meals, and some women are known to make better bread than others (Dufour, pers. comm.). This would be another way of tracking the successful, but it would be neither unconscious nor 'dumb'. What this suggestion leaves out is some account that can substantiate (i) the idea that storing the starch and fibre for longer makes casabe taste better (as suggested by Mercier and Morin 2019) and (ii) some kind of link between better-tasting bread and healthier bread. In Section 4.7, I will try to supplement Henrich's account by making just this case: for the moment, it is important to see that Henrich's other two suggestions also require an account of this sort.

Henrich suggests that attending to the prestigious might be another way in which complex suites of efficacious behaviours accumulate. The rough theory of a hypothesised 'prestige bias' goes like this (Henrich and Gil-White 2001): people tend to copy individuals who are accorded prestige. This bias makes adaptive sense: faced with a choice of whether to copy a model who is prestigious, versus a model who is not, one is better off copying the prestigious model. This is because prestige most likely emerges through a kind of trade. Many valuable skills can only be learned by spending a good amount of time with an adept, but adepts do not necessarily want neophytes to shadow them. The price the skilful demonstrator extracts for allowing learners to profit is the status conferred by deference. Initially, a would-be learner needs to identify who is skilful, and pay them respect in return for learning opportunities. Identifying who is skilful is difficult: a hunter's one-shot success might be a result of good luck, rather than genuine skill. Over time it therefore makes sense to attempt to learn from those who have been accorded prestige, since such individuals have already been identified, based on longer-term success, as skilful by others. This 'prestige bias' pays off, because the prestige accorded to an individual is likely to indicate their mastery of a broad range of valuable techniques.

Azita Chellappoo (2020) has raised many questions for this approach. The most germane here concern whether evidence establishes a broad 'bias' in favour of prestige, as opposed to a flexible tendency to tailor one's learning strategies according to various markers that include the deference shown to practitioners. If one wants to learn the cello, then it makes sense to seek out the best teacher. It is hard to determine who the best teacher is directly. That is because the effects of teaching can take years to manifest themselves; one may not have time to try each teacher for several months before deciding which one to stick with; teachers may not be disposed to demonstrate their skills to anyone who wants to assess them; and so forth. It can make sense, therefore, to see which cellists are most in demand for their teaching. Hence, it can make sense to seek out the most prestigious teachers. Importantly, this way of tracking prestige requires ongoing assessment from students of the quality of teaching offered by the master. In some rare circumstances, and for a short period of time, it may be that all that matters is being seen to have been trained by a particular fêted individual, regardless of how good their teaching is. But this will not be stable: if the teaching offered does not pay off, then students either desert the master, or perhaps these students' own playing will fail to impress others, and eventually prestige will wither. In other words, even when prestige is an important cue, it needs to be deployed in a sensitive and flexible way, and in a manner that cannot float entirely free from an ability to monitor underlying competence.

This intuitive verdict on how one might expect prestige to be deployed has been supported by experimental work. Brand et al. (2020) recruited volunteers to an online quiz with multiple rounds, whose rules allowed the volunteers to choose other recruits to copy when giving answers to general knowledge questions. Brand et al. used the number of times an individual was copied as a measure of their 'prestige', and so the key question for the study was whether volunteers decided who to copy by using information (when it was made available) about who had been copied most often in previous rounds. Their results line up with intuitive expectations: information about the prestige of a potential informant was *not* used when more direct information was available about the participants' actual success in getting answers right in a previous round of the quiz. Prestige information *was* used when information about success was unavailable, but only if the quiz set-up meant that being copied in a previous round was an indicator of getting answers right in an even earlier round. In other words, individuals used prestige in a flexible way depending on whether it seemed to be a good indirect indicator of success.

This rationale for attending to prestige fails to show that it makes sense to learn a general set of skills from prestigious individuals: there is no reason to learn anything much from renowned cello soloists like Steven Isserlis or Sol Gabetta beyond matters related to music. In a different experiment, Brand et al. (2021) found that prestigious individuals were consulted for information about the specific areas in which they had previously shown success, and only about more distantly related matters if learners had no other sources of information relating to those topics. Perhaps unsurprisingly, they suggest their data are best understood as reflecting learners' own varying assumptions – which may or may not be accurate, depending on those learners' background knowledge – about whether an expert in one domain is likely to have expertise that carries over to another. (Consider, for example, that it is not obvious whether Steven Isserlis's expertise as a cellist is a likely indicator of expertise regarding genres of music outside the classical repertoire.)

This flexible interpretation of the role of tracking prestige is consistent with Henrich and Broesch's (2011) detailed work on learning in Fiji. They documented that individuals who were perceived to be successful at fishing would also be sought-after as sources for information about growing yams, but not as sources for information about medicinal plants. They suggest (albeit tentatively) that this can be explained because of perceived links between the more traditionally male domains of yam-growing and fishing. What is more, the fact that yam-growing and fishing are at the top of these communities' lists of skills that 'had to be mastered to be considered a respected member of the community' – whereas knowledge of medical plants is not – potentially gives learners reason to think they may be found together in models: individuals who are well-respected probably have knowledge of both (2011: 1143).

Henrich and Gil-White's earlier illustrative cases of how prestige-bias might emerge are also in line with this contextually sensitive interpretation. For example, they suggest that a one-off observation of a hunter may mislead as to their competence; but they do not claim that it is impossible to evaluate how good someone is at hunting. Indeed, their presentation of prestige bias allows that 'novices are initially better off selecting models who are already favoured by others. Later, after they have accumulated their own long-term samples, they can refine these borrowed judgements. Hunting returns are hard to fake—and if they bring prestige they will be advertised—so information gathering costs are substantially reduced for novices' (2001: 178). In other words, there is a simultaneous process whereby novices make use of prestige as an accessible cue of an individual's worth as a model, while more experienced learners are able to validate whether that prestige is warranted by evaluating actual success. Again, the strong accent Henrich gives to causal opacity makes it hard to see how this can work in the case of manioc processing if, unlike hunting, it is difficult for individuals to determine even in the long term how well a technique reduces cyanogen content. So, if individuals are using prestige, it seems more likely that it is prestige understood as an indicator of being a good bread-maker.

Henrich's final suggestion might be able to square the circle of explaining cumulative adaptation in the context of causal opacity: perhaps learning from the healthy gives a straightforward way of making sure that one copies techniques that, as a matter of fact, work well to reduce cyanogens. Imagine that an individual chances on a slightly better way of detoxifying manioc. The fact that it works means it will contribute positively to this person's health, even if neither originator nor observers realises this. If others copy what the healthy do, then this improvement will spread. Someone else might then make another lucky improvement to the technique, rendering them healthier still. Now they will be the model of choice for others. And so the process repeats, to give rise to the full manioc processing regime.

Henrich and Gil-White conjecture that, 'in small-scale societies lacking division of labour and supporting institutions, arcane endeavours that compromise food production are likely to make practitioners appear unhealthy compared to their neighbours' (2001: 176). It is difficult to transfer this way of thinking to the manioc case. Imagine an earlier state where everyone makes their casabe as fast as possible. Perhaps one individual forgets about it, and only remembers to bake their casabe 48 hours later. On the assumption that it is impossible for individuals to figure out that this benefits health, there is an initial problem in explaining why the individual in question persists in making casabe in this way for long enough that the health benefits would kick in. By hypothesis, it cannot be because they detect that their habit is doing them good. Even assuming that for some reason the individual does persist with a habit that has no appeal for them, it is by no means guaranteed that they will become healthier than others in the population. Their processing technique is good for them, but its beneficial effects may be masked by all sorts of unrelated diseases or injuries that mean they are less healthy than their companions. And even if they are the healthiest in the community, how are individuals observing them supposed to know that it is their casabe-making technique they should be imitating, rather than a host of other idiosyncratic things – walking, singing, wearing their hair – that they may do slightly differently from everyone else? If observers focus on casabe-making – and Dufour reports that women who make good casabe are held in high esteem (pers. comm.) – then it seems likely that the focus of their attention is driven by some hypothesis (albeit potentially a vague one) that waiting 48 hours gives rise to better bread.

4.7 The Sourdough Taste

Henrich's general story of cultural accumulation is intact: I have said nothing to undermine the claims that manioc processing is primarily learnt by women from their mothers, and that the suite of effective steps has gradually accumulated over time. I do suggest this story needs to be supplemented in two ways. I have already suggested that this copying is not 'dumb'. I have not yet added the second supplement, which is some hypothesis to explain why the waiting stage was taken up and proliferated in the first place if its efficacy is unknown. I suggest a plausible account can be fashioned by attending to what Tukanoan people say about their practice.

Moving away, briefly, from the matter of the processing technique, Dufour's original fieldwork on this group makes clear that Tukanoan people understand that their favoured varieties of manioc are particularly toxic (Wilson and Dufour 2002). When pressed on why they favour these varieties, some point to cogent causal rationales: they note, for example, that the sweet varieties are dug up far more often by black agouti. They also know their manioc needs to be processed in order to make it healthy. Dufour indicates that they may have some evidence of the positive detoxifying effects of the 48-hour storage phase. They sometimes feed manioc mash to their chickens, and she reports that, 'One woman told me that she only fed the grated mash to her chickens after it had been stored because feeding freshly grated mash to chickens was dangerous—it could/would kill them' (pers. comm.). Moreover, while Henrich may be right to say that discernible bitterness disappears from the starch and fibre before processing is complete, this does not mean that the additional waiting step has no discernible effect on the taste of the final products made from manioc. Tukanoan people report that when the waiting process is allowed to run its course, the fermentation that ensues gives the casabe a more bread-like open texture, as well as a 'sourdough' taste that they prefer. If, as sometimes happens, casabe is made straight away, then it is much thinner, without the texture akin to a leavened bread that they value (Dufour pers. comm.; see also Mercier and Morin 2019).

This way of telling the story still leaves plenty of unanswered questions about how Tukanoan processing techniques arise and proliferate. In fashioning a response, two things are worth stressing. One is the pride Tukanoan people take in making bread with what is (for them) the requisite texture and taste, and the manner in which making bread of this type is a marker of Tukanoan identity. A second is that many cultures where bitter manioc is consumed do not use processing techniques that are as effective as those of Tukanoan people, and in these places the chronic effects of cyanogen consumption are serious. It is possible, therefore, that Tukanoan people are lucky compared with these other cultures: the value they place on an idiosyncratic form of taste and texture has the fortuitous side effect of making their distinctive way of processing manioc highly efficacious. If they took pride in a different sort of taste and texture, they would have been in the situation of other communities where diseases like goitre are more common. It seems to me, however, that one need not describe this story as one of pure luck, because while the preference for a 'sourdough' taste and an open texture may be especially pronounced among Tukanoan people, it builds on a far more general preference for fermented foods whose origination can be linked to the benefits fermentation bestows. It is not unusual for producers to wait a very long time during the preparation of fermented foods: this may be most obvious when one thinks of wine or cheese production, but French boulangeries often leave their dough to rest for 48 hours before baking, the Swedish fermented fish *surströmming* can take months before it is considered ready to eat, and even the supermarket pizza in my freezer boasts that it is made 'with a 48 hour fermented sourdough for extra flavour' (Waitrose 2024).

Amato and colleagues (2021a, b) conjecture a very longue-durée story of the control of fermentation, which begins with *Australopithecus* (an early ape-like ancestor of modern humans, living between roughly 4 and 2 million years ago). They may have dug up roots and tubers of various kinds, briefly chewed on them, and discarded them finding them unpalatable. It is possible that they returned later to find them tasting better after initial inoculation by microorganisms from their mouths, followed by subsequent fermentation. (There are reports of monkeys – specifically spider monkeys, capuchins and brown lemurs – knocking fruits to the ground and then returning some days later to eat them after fermentation has occurred; and again, bacterial inoculation when they are first bitten into may aid this process.) After this chance discovery, Amato et al. suggest that *Australopithecus* may have been able to initiate fermentation in a more systematic way, perhaps by deliberately chewing, cutting and burying these roots and tubers (2021b: S213). This does not involve ignorance that the techniques are of value – their immediate perceived value is that they make better tasting food – even if it retains the thought that *Australopithecus* has no idea about the chemical nature of fermentation itself, and the form of health benefits it brings.

This story about the control of fermentation presupposes a liking for the taste of fermented foods. The converse, of course, is not the case: a species may like fermented foods without having any ability to control the fermentation process. Frank et al. make the case that a liking for acidic foods – including foods made acidic by fermentation – appears in species when it is adaptive: for example, they provide evidence that wild pigs tend to prefer acid foods, that pigs have a 'strong attraction to fermented baits' (2022: 5), and they conjecture that this attraction may aid pigs in finding safe sources of foods on or under the ground. The consumption of fermented foods is itself associated with various nutritional benefits, ranging from the effects of fermentation on potentially harmful bacteria, to the enhancement of available calories, as well as the breakdown of both toxins and tough outer layers. An attraction to the smell associated with natural fermentation may also have allowed wild animals to find nutritious foods more easily. Frank et al. do not go so far as to suggest that the preference for an acid taste is innate in humans: as they put it, 'both chimpanzees and humans either instinctively prefer acidic foods or readily learn to prefer them' (2022: 6). But regardless of how this preference develops, it does appear to be ancient: they give evidence from molecular studies indicating that the last common ancestor of humans, chimpanzees and gorillas consumed fermented fruits, and they go on to suggest that the preference for this acidic taste would have set the foundations for the later cultural accumulation of control over fermentation.

4.8 From Hidden Hand to Many Hands

Suppose I am right that the story of manioc processing among Tukanoan people, as initially told by Henrich, leaves open some important questions about why they have devised and then persisted with important elements of the processing regime. A reader might wonder what general significance can be drawn from this, beyond the obvious point that it is always possible to add more detail to an explanatory narrative.

It seems to me that there are three broad lessons that can be drawn, and they all signal departures from a model that sees cultural selection as a close analogue to natural selection. They consequently reinforce a key message of this Element, which is that while the study of cultural evolution can certainly make good use of explanatory and investigative tools adapted from mainstream evolutionary theory, their deployment and interpretation needs to be flexible, and to show due deference to the specificities of cultural change. First, neither the story of manioc processing, nor Henrich's own general schema for cumulative cultural adaptation, is a thoroughgoing hidden-hand explanation. It is more like a 'many hands' explanation. Cultural evolution allows individuals to benefit from the investigative labour of their predecessors; in that sense, cultural evolution is indeed smarter than any single individual. But this does not make cultural evolution 'dumb': it allows the accumulation and pooling of these many earlier instances of insight.

Second, and relatedly, this account illustrates one way in which the sort of quasi-selectionist explanations that I have drawn attention to in Sections 1 and 2 face constraints that are unlike those in the organic context. For while broadly selectionist narratives certainly do not require individual agents to fully grasp the beneficial effects of their techniques, there does need to be some plausible account that explains what value they see in them, which can explain why they bother to adopt and maintain those techniques.

Finally, the specific way in which I have filled out the details of such an account draws on the preference for 'sourdough-tasting' casabe, which seems to be a culturally accentuated instance of a much more general tendency to seek out fermented foods of many kinds. This preference appears to be ancient, extending deep into the past of the *Homo* genus. It is, in the Parisian parlance, a particularly broad-ranging 'factor of attraction'. This does not entail that the preference for fermented foods is innate, but 'factors of attraction' need not be innate. Whatever the true developmental story for this taste preference, its origins seem to be explained by the general nutritional benefits of fermentation, and the preference itself helps to explain the subsequent control and elaboration of fermentation technologies. This story exemplifies, I hope, this Element's advocacy of a combination of the Paris School's work on cultural attraction with the California School's quasi-selectionist focus on cultural accumulation, in a manner that acknowledges the flexibility and context-sensitivity of learning.

References

Acerbi, A. and A. Mesoudi (2015) 'If We're All Cultural Darwinians What's the Fuss about? Clarifying Recent Disagreements in the Field of Cultural Evolution' *Biology and Philosophy* 30: 481–503.

Amato, K., O. M. Chaves, E. Mallott et al. (2021a) 'Fermented Food Consumption in Wild Nonhuman Primates and Its Ecological Drivers' *American Journal of Biological Anthropology* 175: 513–530.

Amato, K., E. Mallot, P. D'Almeida Maia and M. L. S. Sardaro (2021b) 'Predigestion as an Evolutionary Impetus for Human Use of Fermented Food' *Current Anthropology* 62: S207–S219.

Amundson, R. (1989) 'The Trials and Tribulations of Selectionist Explanations' in K. Hahlweg and C. Hooker (eds.) *Issues in Evolutionary Epistemology*. Albany: SUNY Press, pp. 413–432.

Baravalle, L. and V. Luque (2022) 'Towards a Pricean Foundation for Cultural Evolutionary Theory' *Theoria* 37: 209–232. https://doi.org/10.1387/theoria.21940

Basalla, G. (1988) *The Evolution of Technology*. Cambridge: Cambridge University Press.

Birch, J. (2013) *Kin Selection: A Philosophical Analysis*. PhD Dissertation, University of Cambridge.

Birch, J. (2017) *The Philosophy of Social Evolution*. Oxford: Oxford University Press.

Birch, J. and C. Heyes (2021) 'The Cultural Evolution of Cultural Evolution' *Philosophical Transactions of the Royal Society B* 376: 20200051. http://doi.org/10.1098/rstb.2020.0051

Boyd, R. and P. Richerson (1985) *Culture and the Evolutionary Process*. Chicago: University of Chicago Press.

Boyd, R. and P. Richerson (1996) 'Why Culture Is Common, but Cultural Evolution Is Rare' *Proceedings of the British Academy* 88: 77–93.

Boyd, R. and P. Richerson (2000) 'Memes: Universal Acid or a Better Mousetrap?' in R. Aunger (ed.) *Darwinizing Culture*. Oxford: Oxford University Press, pp. 143–162.

Brand, C., S. Heap, T. Morgan and A. Mesoudi (2020) 'The Emergence and Adaptive Use of Prestige in an Online Social Learning Task' *Scientific Reports* 10: 12095.

Brand, C., A. Mesoudi and T. Morgan (2021) 'Trusting the Experts: The Domain-Specificity of Prestige-Biased Social Learning' *PLoS ONE* 16: e0255346. https://doi.org/10.1371/journal.pone.0255346

Buskell, A. (2017a) 'What Are Cultural Attractors?' *Biology and Philosophy* 32: 377–394.

Buskell, A. (2017b) 'Cultural Attractor Theory and Explanation' *Philosophy, Theory, and Practice in Biology* 9: 13.

Buskell, A. (2022) 'Cumulative Culture and Complex Cultural Traditions' *Mind and Language* 37: 284–303.

Caldwell, C. and A. Millen (2009) 'Social Learning Mechanisms and Cumulative Cultural Evolution: Is Imitation Necessary?' *Psychological Science* 20: 1478–1483.

Campbell, D. (1960) 'Blind Variation and Selective Retentions in Creative Thought as in Other Knowledge Processes' *Psychological Review* 67: 380–400.

Campbell, D. (1974) 'Evolutionary Epistemology' in P. A. Schilpp (ed.) *The Philosophy of Karl Popper*. La Salle: Open Court, pp. 413–463.

Carlson, W. B. (2000) 'Invention and Evolution: The Case of Edison's Sketches of the Telephone' in J. Ziman (ed.) *Technological Innovation as an Evolutionary Process*. Cambridge: Cambridge University Press, pp. 137–158.

Cavalli-Sforza, L. and M. Feldman (1973) 'Cultural versus Biological Inheritance: Phenotypic Transmission from Parents to Children' *American Journal of Human Genetics* 25: 618–637.

Cavalli-Sforza, L. and M. Feldman (1981) *Cultural Transmission and Evolution: A Quantitative Approach*. Princeton: Princeton University Press.

Charbonneau, M. (2016) 'Evo-Devo and Culture' in L. Nuño de la Rosa and G. Müller (eds.) *Evolutionary Developmental Biology*. Switzerland: Springer, pp. 1235–1248.

Charbonneau, M. (2020) 'Understanding Cultural Fidelity' *The British Journal for the Philosophy of Science* 71(4): 1209–1233.

Charbonneau, M. and P. Bourrat (2021) 'Fidelity and the Grain Problem in Cultural Evolution' *Synthese* 199: 5815–5836.

Chellappoo, A. (2020) 'Rethinking Prestige Bias' *Synthese* 198: 8191–8212. https://doi.org/10.1007/s11229-020-02565-8

Chellappoo, A. (2022) 'When Can Cultural Selection Explain Adaptation?' *Biology and Philosophy* 37: 2.

Claidière, N., T. Scott-Phillips and D. Sperber (2014) 'How Darwinian Is Cultural Evolution?' *Philosophical Transactions of the Royal Society B* 369: 20130368. https://doi.org/10.1098/rstb.2013.0368

Cobb, M. and N. Comfort (2023) 'What Watson and Crick Really Took from Franklin' *Nature* 616: 657–660.

Cziko, G. (1997) *Without Miracles: Universal Selection Theory and the Second Darwinian Revolution*. Cambridge, MA: MIT Press.

Darwin, C. (1859) *On the Origin of Species*. London: John Murray.

Darwin, C. (1871) *The Descent of Man and Selection in Relation to Sex*. London: John Murray.

Dawkins, R. (1976) *The Selfish Gene*. Oxford: Oxford University Press.

Dawkins, R. (1983) 'Universal Darwinism' in D. S. Bendall (ed.) *Evolution: From Molecules to Man*. Cambridge: Cambridge University Press, pp. 403–425.

Dennett, D. (1995) *Darwin's Dangerous Idea*. New York: Simon and Schuster.

Dennett, D. (2017) *From Bacteria to Bach and Back*. New York: Norton.

Derex, M. (2022) 'Human Cumulative Culture and the Exploitation of Natural Phenomena' *Philosophical Transactions of the Royal Society B* 377: 1–10.

Driscoll, C. (2011) 'Fatal Attraction? Why Sperber's Attractors Do Not Prevent Cumulative Cultural Evolution' *British Journal for the Philosophy of Science* 62: 301–322.

Dufour, D. (1994) 'Cassava in Amazonia: Lessons in Utilization and Safety from Native Peoples' *Acta Horticulturae* 375: 175–182.

Dufour, D. (1995) 'A Closer Look at the Nutritional Implications of Bitter Cassava Use' in L. Sponsel (ed.) *Indigenous Peoples and the Future of Amazonia: An Ecological Anthropology of an Endangered World*. Tucson: The University of Arizona Press, pp. 149–165.

Dufour, D. (2006) 'Biocultural Approaches in Human Biology' *American Journal of Human Biology* 18: 1–9.

El Mouden, C., J.-B. André, O. Morin and D. Nettle (2013) 'Cultural Transmission and the Evolution of Human Behaviour: A General Approach Based on the Price Equation' *Journal of Evolutionary Biology* 27: 231–241.

Enquist, M., P. Strimling, K. Eriksson, K. Laland and J. Sjostrand (2010) 'One Cultural Parent Makes no Culture' *Animal Behaviour* 79: 1353–1362.

Food and Agricultural Organization of the United Nations (2013) *Save and Grow: Cassava: A Guide to Sustainable Production Intensification*. Rome: Food and Agricultural Organization of the United Nations.

Frank, H., K. Amato, M. Trautwein et al. (2022) 'The Evolution of Sour Taste' *Proceedings of the Royal Society B*: 289: 20211918.

Frank, S. (1995) 'George Price's Contributions to Evolutionary Genetics' *Journal of Theoretical Biology* 175: 373–388.

Gerhart, J. and M. Kirschner (2007) 'The Theory of Facilitated Variation' *PNAS* 104: 8582–8589.

Godfrey-Smith, P. (2009) *Darwinian Populations and Natural Selection.* Oxford: Oxford University Press.

Godfrey-Smith, P. (2012) 'Darwinism and Cultural Change' *Philosophical Transactions of the Royal Society B* 367: 2160–2170.

Gray, R. D., S. J. Greenhill and R. M. Ross (2007) 'The Pleasures and Perils of Darwinizing Culture (With Phylogenies)' *Biological Theory* 2(4): 360–375.

Helanterä, H. and T. Uller (2010) 'The Price Equation and Extended Inheritance' *Philosophy and Theory in Biology* 2(201306): 1–17.

Helanterä, H. and T. Uller (2020) 'Different Perspectives on Non-genetic Inheritance Illustrate the Versatile Utility of the Price Equation in Evolutionary Biology' *Philosophical Transactions of the Royal Society B* 375: 20190366.

Henrich, J. (2004) 'Cultural Group Selection, Coevolutionary Processes and Large-Scale Cooperation' *Journal of Economic Behavior and Organization* 53: 3–35.

Henrich, J. (2016) *The Secret of Our Success.* Princeton: Princeton University Press.

Henrich, J. and R. Boyd (1998) 'The Evolution of Conformist Transmission and the Emergence of Between-Group Differences' *Evolution and Human Behavior* 19: 215–241.

Henrich, J. and R. Boyd (2002) 'On Modeling Cognition and Culture: Why Cultural Evolution Does Not Require Replication of Representations' *Journal of Cognition and Culture* 2: 87–112.

Henrich, J. and J. Broesch (2011) 'On the Nature of Cultural Transmission Networks: Evidence from Fijian Villages for Adaptive Learning Biases' *Philosophical Transactions of the Royal Society B* 366: 1139–1148.

Henrich, J. and F. Gil-White (2001) 'The Evolution of Prestige: Freely Conferred Deference as a Mechanism for Enhancing the Benefits of Cultural Transmission' *Evolution and Human Behavior* 22: 165–196.

Henrich, J., R. Boyd and P. Richerson (2008) 'Five Misunderstandings about Cultural Evolution' *Human Nature* 19: 119–137.

Heyes, C. (2018) *Cognitive Gadgets: The Cultural Evolution of Thinking.* Cambridge, MA: Harvard University Press.

Hoehl, S., S. Keupp, H. Schleihauf, et al. (2019) '"Over-Imitation": A Review and Appraisal of a Decade of Research' *Developmental Review* 51: 90–108.

Horner, V. and A. Whiten (2005) 'Causal Knowledge and Imitation/Emulation Switching in Chimpanzees (*Pan troglodytes*) and Children (*Homo sapiens*)' *Animal Cognition* 8: 164–181.

Ingold, T. (2022) 'Evolution without Inheritance: Steps to an Ecology of Learning' *Current Anthropology* 63: S32–S55.

Jablonka, E. and M. Lamb (2014) *Evolution in Four Dimensions*. Cambridge, MA: MIT Press.

Kerr, B. and P. Godfrey-Smith (2009) 'Generalization of the Price Equation for Evolutionary Change' *Evolution* 63: 531–536.

Kronfeldner, M. (2007) 'Is Cultural Evolution Lamarckian?' *Biology and Philosophy* 22: 493–512.

Laland, K. (2018) *Darwin's Unfinished Symphony: How Culture Made the Human Mind*. New Haven: Princeton University Press.

Laland, K., T. Uller, M. Feldman et al. (2015) 'The Extended Evolutionary Synthesis: Its Structure, Assumptions and Predictions' *Proceedings of the Royal Society* B 282: 1–14.

Laland, K. N., J. Odling-Smee and M. Feldman (2000) 'Niche Construction, Biological Evolution, and Cultural Change' *Behavioral and Brain Sciences* 23(1): 131–175.

Lehmann, L. and M. Feldman (2008) 'The Co-evolution of Culturally Inherited Altruistic Helping and Cultural Transmission under Random Group Formation' *Theoretical Population Biology* 73: 506–516.

Lewens, T. (2004) *Organisms and Artifacts: Design in Nature and Elsewhere*. Cambridge, MA: MIT Press.

Lewens, T. (2007) *Darwin*. London: Routledge.

Lewens, T. (2009) 'What's Wrong with Typological Thinking?' *Philosophy of Science* 79: 355–371.

Lewens, T. (2015) *Cultural Evolution: Conceptual Challenges*. Oxford: Oxford University Press.

Lewens, T. (2020) 'How Can Conceptual Analysis Contribute to Scientific Practice? The Case of Cultural Evolution' in T. Uller and K. Kampourakis (eds.) *Philosophy of Science for Biologists*. Cambridge: Cambridge University Press, pp. 146–161.

Lewens, T. (2022) 'Philosophy of Cultural Evolution' in J. Tehrani, J. Kendall and R. Kendall (eds.) *The Oxford Handbook of Cultural Evolution*. Oxford: Oxford University Press. https://doi.org/10.1093/oxfordhb/9780198869252.013.10.

Lewens, T. (2023) 'Equations at an Exhibition' in A. du Crest, M. Valković, A. Ariew, H. Desmond, P. Huneman and T. Reydon (eds.) *Evolutionary Thinking Across Disciplines*. Switzerland: Springer, pp. 353–373.

Lewens, T. and A. Buskell (2023) 'Cultural Evolution' in E. N. Zalta and U. Nodelman (eds.) *Stanford Encyclopedia of Philosophy* (Summer 2023 Edition). https://plato.stanford.edu/archives/sum2023/entries/evolution-cultural/

Lewis, H. and K. Laland (2012) 'Transmission Fidelity Is the Key to the Build-Up of Cumulative Culture' *Philosophical Transactions of the Royal Society B* 367: 2171–2180.

Mercier, H. and O. Morin (2019) 'Blind Imitation or a Matter of Taste?' Blog Post: http://cognitionandculture.net/blogs/hugo-mercier/a-matter-of-taste/. Accessed 8th January 2024.

Mesoudi, A. (2008) 'Foresight in Cultural Evolution' *Biology and Philosophy* 23: 243–255.

Mesoudi, A. (2011) *Cultural Evolution: How Darwinian Theory Can Explain Human Culture and Synthesize the Social Sciences*. Chicago: University of Chicago Press.

Mesoudi, A. (2021) 'Blind and Incremental or Directed and Disruptive? On the Nature of Novel Variation in Human Cultural Evolution' *American Philosophical Quarterly* 58: 7–20.

Mesoudi, A. and A. Thornton (2018) 'What Is Cumulative Cultural Evolution?' *Proceedings of the Royal Society B* 285: 20180712.

Mesoudi, A., A. Whiten and K. Laland (2004) 'Perspective: Is Human Cultural Evolution Darwinian? Evidence Reviewed from the Perspective of *The Origin of Species*' *Evolution* 58(1): 1–11.

Mesoudi, A., A. Whiten and K. Laland (2006) 'Towards a Unified Science of Cultural Evolution' *Behavioral and Brain Sciences* 29: 329–347.

Morin, O. (2013) 'How Portraits Turned Their Eyes upon Us: Visual Preferences and Demographic Change in Cultural Evolution' *Evolution and Human Behaviour* 34: 222–229. https://doi.org/10.1016/j.evolhumbehav.2013.01.004

Morin, O. (2016) *How Traditions Live and Die*. Oxford: Oxford University Press.

Neander, K. (1995) 'Pruning the Tree of Life' *British Journal for the Philosophy of Science* 46: 59–80.

Nettle, D. (2020) 'Selection, Adaptation, Inheritance and Design in Human Culture: The View from the Price Equation' *Philosophical Transactions of the Royal Society B* 375: 20190358.

Newman, S. and G. Müller (2001) 'Epigenetic Mechanisms of Character Origination' in G. Wagner (ed.) *The Character Concept in Evolutionary Biology*. San Diego: Academic Press, 561–581.

Odling-Smee, J., K. Laland and M. Feldman (2003) *Niche Construction: The Neglected Process in Evolution*. Princeton: Princeton University Press.

Okasha, S. (2006) *Evolution and the Levels of Selection*. Oxford: Oxford University Press.

Okasha, S. and J. Otsuka (2020) 'The Price Equation and the Causal Analysis of Evolutionary Change' *Philosophical Transactions of the Royal Society B* 375: 20190365.

Pinker, S. (1997) *How the Mind Works*. London: Allen Lane.

Price, G. (1970) 'Selection and Covariance' *Nature* 227: 520–521.

Price, G. (1972) 'Extension of Covariance Selection Mathematics' *Annals of Human Genetics* 35: 485–490.

Price, G. (1995) 'The Nature of Selection' *Journal of Theoretical Biology* 175: 389–396.

Ramsey, G. and A. De Block (2017) 'Is Cultural Fitness Hopelessly Confused?' *The British Journal for the Philosophy of Science* 68: 305–328.

Rice, S. (2004) *Evolutionary Theory*. Sunderland: Sinauer.

Richerson, P. and R. Boyd (2005) *Not by Genes Alone: How Culture Transformed Human Evolution*. Chicago: University of Chicago Press.

Richerson, P., R. Baldini, A. V. Bell, et al. (2016) 'Cultural Group Selection Plays an Essential Role in Explaining Human Cooperation: A Sketch of the Evidence' *Behavioral and Brain Sciences* 39: e30. https://doi.org/10.1017/S0140525X1400106X

Russ, M. (1992) *Musorgsky: Pictures at an Exhibition*. Cambridge: Cambridge University Press.

Salazar-Ciudad, I. and J. Jernvall (2010) 'A Computational Model of Teeth and the Developmental Origins of Morphological Variation' *Nature* 464: 583–586.

Scott-Phillips, T., S. Blancke and C. Heintz (2018) 'Four Misunderstandings about Cultural Attraction' *Evolutionary Anthropology* 27: 162–173.

Sperber, D. (1996) *Explaining Culture: A Naturalistic Approach*. Oxford: Blackwell.

Sperber, D. (2000) 'An Objection to the Memetic Approach to Culture' in R. Aunger (ed.) *Darwinizing Culture: The Status of Memetics as a Science*. Oxford: Oxford University Press, pp. 163–174.

Sperber, D. (2001) 'Conceptual Tools for a Natural Science of Society and Culture' *Proceedings of the British Academy* 111: 297–317.

Sperber, D. and N. Claidière (2008) 'Defining and Explaining Culture' *Biology and Philosophy* 23: 283–292.

Sterelny, K. (2012) *The Evolved Apprentice: How Evolution Made Humans Unique*. Cambridge, MA: MIT Press.

Sterelny, K. (2017) 'Cultural Evolution in Paris and California' *Studies in History and Philosophy of Biological and Biomedical Sciences* 62: 42–50.

Sterelny, K. (2021) *The Pleistocene Social Contract: Culture and Cooperation in Human Evolution*. Oxford: Oxford University Press.

Tennie, C., J. Call and M. Tomasello (2009) 'Ratcheting Up the Ratchet: On the Evolution of Cumulative Culture' *Philosophical Transactions of the Royal Society B: Biological Sciences* 364(1528): 2405–2415.

Tomasello, M. (1999) *The Cultural Origins of Human Cognition*. Cambridge, MA: Harvard University Press.

Tomasello, M., A. Kruger and H. Ratner (1993) 'Cultural Learning' *Behavioral and Brain Sciences* 16: 495–552.

Uller, T., A. Moczek, R. Watson, P. Brakefield and K. Laland (2018) 'Developmental Bias and Evolution: A Regulatory Network Perspective' *Genetics* 209: 949–966.

Waitrose (2024) *No. 1 Roasted Vegetable and Pesto Sourdough Pizza* (Food Packaging). Bracknell: Waitrose.

Wilson, W. and D. Dufour (2002) 'Why "Bitter" Cassava?' *Economic Botany* 56: 49–57.

Acknowledgements

I have had a huge amount of help from generous friends, colleagues and family members over the years, all of whom have contributed in one way or another to this Element. For comments and advice I am especially grateful to Anna Alexandrova, Jonathan Birch, Andrew Buskell, Azita Chellappoo, Heidi Colleran, Darna Dufour, Harriet Fagerberg, Laurel Fogarty, Steve Frank, Emma Gilby, Marta Halina, Celia Heyes, Philippe Huneman, Kevin Lala, Maël Lemoine, Mathilde Lequin, Julieta Macome, Hugo Mercier, Alex Mesoudi, Olivier Morin, Arsham Nejad Kourki, Daniel Nettle, Samir Okasha, Thomas Pradeu, Zeshan Qureshi, Grant Ramsey, Raphael Scholl, Philipp Spillmann, Kim Sterelny, Richard Watson, Cassandra Yang, Jolie Zhou and audiences at the University of Bordeaux and the University of Hannover.

I am grateful to Springer for permission to reuse material from Lewens, T. (2023) 'Equations at an Exhibition' in A. du Crest et al. (eds.) *Evolutionary Thinking Across Disciplines*: Springer. Figures 1 and 2 were kindly provided by The Thomas A. Edison Papers at Rutgers University under a creative commons licence.

This work is dedicated to the memory of Peter Lipton.

Grant Ramsey

KU Leuven

Grant Ramsey is a BOFZAP research professor at the Institute of Philosophy, KU Leuven, Belgium. His work centers on philosophical problems at the foundation of evolutionary biology. He has been awarded the Popper Prize twice for his work in this area. He also publishes in the philosophy of animal behavior, human nature and the moral emotions. He runs the Ramsey Lab (theramseylab.org), a highly collaborative research group focused on issues in the philosophy of the life sciences.

Michael Ruse

Florida State University

Michael Ruse is the Lucyle T. Werkmeister Professor of Philosophy and the Director of the Program in the History and Philosophy of Science at Florida State University. He is Professor Emeritus at the University of Guelph, in Ontario, Canada. He is a former Guggenheim fellow and Gifford lecturer. He is the author or editor of over sixty books, most recently *Darwinism as Religion: What Literature Tells Us about Evolution*; *On Purpose*; *The Problem of War: Darwinism, Christianity, and their Battle to Understand Human Conflict*; and *A Meaning to Life*.

About the Series

This Cambridge Elements series provides concise and structured introductions to all of the central topics in the philosophy of biology. Contributors to the series are cutting-edge researchers who offer balanced, comprehensive coverage of multiple perspectives, while also developing new ideas and arguments from a unique viewpoint.

Cambridge Elements ☰

Philosophy of Biology

Elements in the Series

Philosophy of Developmental Biology
Marcel Weber

Evolution, Morality and the Fabric of Society
R. Paul Thompson

Structure and Function
Rose Novick

Hylomorphism
William M. R. Simpson

Biological Individuality
Alison K. McConwell

Human Nature
Grant Ramsey

Ecological Complexity
Alkistis Elliott-Graves

Units of Selection
Javier Suárez and Elisabeth A. Lloyd

Evolution and Development: Conceptual Issues
Alan C. Love

Inclusive Fitness and Kin Selection
Hannah Rubin

Animal Models of Human Disease
Sara Green

Cultural Selection
Tim Lewens

A full series listing is available at: www.cambridge.org/EPBY

Printed in the United States
by Baker & Taylor Publisher Services